Switching Mode Circuit Analysis and Design: Innovative Methodology by Novel Solitary Electromagnetic Wave Theory

First Edition

Authored By

Hirokazu Tohya

CEO President
ICAST, Inc.
Hachioji-shi
Tokyo
Japan

Short Summary of My eBook

This eBook presents the novel theory and technologies for the design of the switching mode circuit.

BIOGRAPHY OF HIROKAZU TOHYA

He received a bachelor's degree at Kagoshima University in 1968. He received a doctoral degree at Kyushu University in 1999. He belonged to the computer division of NEC during 1968-1992 and belonged to NEC laboratories until 2005. He established ICAST, Inc. in 2005. Currently he is CEO and President of ICAST, Inc.

CONTENTS

FOREWORD

The electric or electronics engineers are addressing the difficult problems at the development of the equipment and the systems. Among them, the EMC problem of SMC is considered to be one of the most difficult problems. EMC consists of the electromagnetic susceptibility (EMS) or Immunity and EMI. All SMC consist of semiconductor devices and the transmission line. The transmission line consists of the insulator and the wire, which is called the interconnect in the case of LSI and is called the trace in the case of PCB. The semiconductor devices change the electromagnetic field on SMC. According to the electromagnetism, the changed electric field (EF) or the changing magnetic field (MF) forms the electromagnetic wave (EMW). The EMW which is generated by the semiconductor device travels in the insulator of the transmission line at light speed approximately. The reason is the following; the length of the interconnect which forms antennas is quite short, and the transistors in LSI are broken down easily by it when it is being handled. In addition, the timing control is the most important for the interconnect designs however it cannot be analyzed by the conventional EMW theory in the electromagnetism.

Many types of the instruments, CAD tools, and the simulators have been used for the circuit design and EMC design of SMC. The time domain types have been used for the circuit design of SMC usually. Symmetrically, the frequency domain types have been used for the EMC design of SMC usually. Because timing design of the signals is the main objective for the circuit design of SMC, and to minimize the electromagnetic interference with the broadcasting service is the main objective of the EMC design of SMC. As the result, the conflicts between the circuit designer and the EMC designer may develop sometimes.

This eBook is focused to improvement of both circuit design and EMC design of SMC. This eBook will present the solution or the guideline which is accepted willingly by the engineers of the circuit design and the EMC design of SMC because the presented novel theory and technologies are discussed based on the conventional EMW theory and SEMW theory. SMC is defined as the electronics circuit which works by the large amplitude motion of the semiconductor such as the transistor. Also the electric circuit which works by the switching motion of the electromechanical component such as the relay is SMC. SMC is being used in almost all electric and electronics equipment today. In contrast, the analog circuit is defined as the electric circuit which is being applied the small amplitude motion of the semiconductor such as the transistor or the dielectric and magnetic materials. The analog circuit is used in a limited field today.

Hirokazu Tohya

ICAST, Inc.
Hachioji-shi, Tokyo192-0912
Japan

PREFACE

I joined NEC in 1968 and was assigned to the circuit development department of the computer division. My first work was the circuit design of the switching mode power supply (SMPS) for the mainframe of the computer system. My job area was expanded to development of the design rule and to standardization of all over the electric and electronics parts for the non-logic circuit. Non-logic circuit was categorized to except the gate-array which forms the logic circuit of the mainframe. The experiment of SMPS was considered to be dangerousness and no one wanted to design. Therefore, my development of the power supply system (PSS) including the circuit design of SMPS was continued. The electric current of the mainframe and the supercomputer was reached to several thousand amperes. The critical problem of PSS was the heat dissipation and the electromagnetic interference (EMI) against the logic circuit. In 1992, the full-scale R&D of EMC became necessary hastily. I was nominated and transferred to the chief managing researcher of the established EMC engineering center in NEC Laboratories. I had believed that the switching mode circuit (SMC) including the digital circuit and SMPS circuit will be improved greatly if the EMC problem is solved. However the deep R&D was difficult because the actual fruits in every year were required strongly. Therefore, his knowledge of the science concerning EMC was not deepened enough. However I could understand the actual status of technologies of the electronics and electrics widely and the basic knowledge about how to approach the science as well as R&D could be learned. During this time, I completed PhD course and received the PhD. degree from Kyushu University, lectured in Kyushu University and Tohoku University for one year each, and executed several national projects by the cooperation of his colleague in NEC and of the supporters in the industrial society and academia. These are considered to be my best harvest in NEC Laboratories. I had decided to continue R&D for solving EMC problem of SMC after his retirement and to try the commercialization of the fruit of R&D. I established ICAST which is the abbreviation of the innovative circuit and system technologies in 2005. The solitary electromagnetic (SEMW) theory was advocated and the novel technologies including the low impedance lossy line (LILL) and the matched impedance lossy line (MILL) were invented based on the SEMW theory.

Two purposes exist in this eBook. The first is to be validated the SEMW theory by the academia and the industrial society. The second is the contribution to the growth of the industry in the world by the technologies of LILL and MILL.

Hirokazu Tohya
ICAST, Inc.
Hachioji-shi, Tokyo192-0912
Japan
E-mail: h-toya@icastech.jp
http://www.icastech.jp

ACKNOWLEDGEMENTS

2.1, Chapter 6, 7.1, and 7.2 include the part of the research report of "Development of Simulation Technologies about Behavior of LSI Circuit (in Japanese)" which was the project commissioned by the Japanese new energy and industrial technology development organization (NEDO). This project was executed in NEC Laboratories between in 2000 and 2001. The member of this project was as following; author as the head, Shiro Yoshida, Kaoru Narita, Hiroshi Abe, Yasushi Kinoshita, Takashi Nakano, Hideto Matsuyama, Naoki Kobayashi, Noriaki Ando, Kushta Taras, Toru Mori, Ichiro Hirata, and Yuichi Yoshida. Shiro Yoshida, Kaoru Narita, Hiroshi Abe, and Toru Mori were the principal researcher of NEC Laboratories at that time. Yasushi Kinoshita, Takashi Nakano, and Hideto Matsuyama were the chief researcher of NEC Laboratories at that time. Naoki Kobayashi, Noriaki Ando, and Kushta Taras were the researcher of NEC Laboratories at that time. Ichiro Hirata was the principal engineer of NEC Jisso and Production Technologies Research (JPTR) Laboratories at that time. Yuichi Yoshida was the chief engineer of NEC JPTR Laboratories at that time. **3.1** shows a part of the result of the collaborative research with Koichiro Masuda and Tomotsugu Arai who were the chief researcher of NEC Laboratories at that time. **Chapter 4** shows a part of the result of the collaborative research with Shiro Yoshida who was the chief engineer of NEC Laboratories at that time. **5.1** is owed to Norio Masuda, Naoya Tamaki, Masahiro Yamaguchi, and Kenichi Arai. Norio Masuda was the chief researcher of NEC Laboratories at that time. Naoya Tamaki was the researcher of NEC Laboratories at that time. Masahiro Yamaguchi was the assistant professor of Tohoku University at that time. Kenichi Arai was the professor of Tohoku University at that time. **5.2** includes the part of the research report of "Semiconductor EMC joint project (in Japanese) which was executed by author as the leader during from 1997 to 2003. The member of this project was consisted of many researchers and engineers of NEC Laboratories and NEC Electronics. **7.3** shows a part of the result of the collaborative research with Koichiro Masuda and Tomotsugu Arai. Koichiro Masuda was the chief researcher of NEC Laboratories at that time. Tomotsugu Arai was the chief engineer of NEC Laboratories at that time. **Chapter 8** includes the part of the research report of "Actualization Development of the low impedance line component (in Japanese)" which was the project commissioned by NEDO and was in NEC JPTR Laboratories dureing in 2002 and 2004. The member of this project was as following; author as the head, Shiro Yoshida, Hiroshi Abe, Kaoru Narita, Takashi Nakano, Hideki Shimizu, Koichiro Masuda, Yoshimasa Wakabayashi, Kushta Taras, Masatoshi Ogawa, Ken Morishita, Satoshi Hukuhara, Akira Onozawa, Manabu Kusumoto, and Kenji Usui. Shiro Yoshida, Hiroshi Abe, Kaoru Narita, Takashi Nakano, and Kushta Taras were the principal researchers of NEC JPTR Laboratories, Hideki Shimizu, Koichiro Masuda, and Yoshimasa Wakabayashi were the chief researchers of NEC JPTR Laboratories at that time. Masatoshi Ogawa, Ken Morishita, Satoshi Hukuhara, Akira Onozawa, Manabu Kusumoto were the researchers of NEC JPTR Laboratories at that time. Kenji Usui was the chief engineer of NEC JPTR Laboratories. **Chapter 10** shows a part of the result of the collaborative research with Heraeus in Germany, Kohzan Corporation, JAPAN CAPACITOR INDUSTRIAL CO., LTD., NEC Personal Products Ltd, VCCI Council, and Mr. Hidetsune Kurokawa who had retired from NEC. **Figure 4** in **Chapter 12** and **12.4** shows a part of the result of the collaborative research with the server platform division of NEC System Technologies, Ltd. **Chapter 12, Chapter 13, and Chapter 14** were collaborated with Noritaka Toya who is the board member of ICAST. Mathcad© was used for the calculation unless otherwise noted.

INTRODUCTION

The electromagnetic susceptibility (EMS) or the immunity and the electromagnetic interference (EMI) are called the electromagnetic compatibility (EMC) as a whole. These are the critical problems in the practical use of the switching mode circuit (SMC) as well as the analog circuit. All SMC consist of semiconductor devices including LSI and the interconnects or the wires. EMS has been paid attention by LSI manufactures, however EMI has not been considered enough because the length of the interconnects which forms antennas is relatively short. EMI is analyzed in accordance with the electromagnetism. The conventional electromagnetism is being little understood by the designers and researchers of LSI. In addition, the timing skew which is the most important item of SMC as well as LSI cannot be analyzed by the conventional electromagnetism.

SMC is defined as the electric circuit which is applied the large amplitude motion of the semiconductor such as the transistor or the switching motion of the electromechanical component such as the relay. SMC is being used in almost all electric and electronics equipment today. In contrast, the analog circuit is defined as the electric circuit which is being applied the small amplitude motion of the semiconductor such as the transistor or the dielectric and magnetic materials. The analog circuit is used in a limited field today.

SMC consists of the switching device, the power line, and the signal line. The signal voltage or current is formed by the switching device such as the transistor, but the SMC cannot work without the power line and signal line. SMC includes the digital circuit, the switching mode power supply (SMPS), and the electrical inverter circuit. These circuits have been being designed and analyzed by the AC circuit theory. It has been believed that SMC is a kind of the analog circuit and the signal voltage of SMC is the kind of the distorted wave. Fourier Transform is being used for analyzing the signal voltage of SMC. According to the idea of Fourier transform, the signal voltage of SMC consists of many harmonic continuous waves and the stationary state voltage depends on the duty cycle. However, some problems occur when this idea is applied to SMC. The 1st problem is that the analysis of the timing skew becomes impossible. The 2nd

problem is that the state of SMC does not match the physics. The electric field of SMC does not change except the switching moment if there are no reflection waves. The third problem is that the stationary state depending on the duty-cycle. The on-state and the off-state exist on SMC explicitly and both states are the stationary states.

The conventional circuit design technologies of SMC are overviewed in Chapter **1**. Many of the information in the overview of LSI about these were drawn from the web site of the international technologies of roadmap for semiconductor (ITRS). The availability of the repeater is also presented. The overview of PCB is presented with the the measured and simulated result. The overview of SMPS is also presented. EMI on SMC is the most critical problem at the designing and manufacturing in the modern electric and electronics products. The study result for suppressing EMI of SMC in accordance with the conventional theory and engineering is presented in Chapter **2** and **3**. The magnet probe was developed for detecting the magnetic field on the power line. It is confirmed that the magnetic probe is useful to suppress the EMI by LSI on PCB by the evaluation of the typical LSI. It was standardized in IEC as the magnetic probe method (MP Method). The result of the feasibility study for suppressing EMI of LSI and is presented. The study result of improved EMI Suppressing techniques of PCB is preseted. It was clarified that using the capacitor becomes the obstacle for enhancing the decoupling performance because the impedance increases at the high-frequency. The result of the feasibility study for reconfiguring SMC to the quasi stationary state closed circuit (QSCC) is presented in Chapter **4**. The limitation of the signal line was considered for suppressing the radiated emission and reducing the crosstalk by the measurement and simulation. QSCC of the power line was formed by using the slender power line and the capacitors. The radiated emission from the prototyped PCB was improved 10dB approximately. This value is considered to depend on the decoupling circuit of the power distribution network (PDN) in spite of forming QSCC of the signal line. The discussion result about the suitable decoupling component as the alternative of the capacitor is presented in Chapter **5**. The review of the characteristics of the decoupling capacitor and the result of the feasibility study about the low-impedance line structure component (LILC) was presented. The prototyped result of the chip ceramic LILC, the cylinder ceramic LILC, and the solid aluiminum LILC are presented. The most suitable solid aluiminum LILC was improved. The result of the feasibility study of the solid aluminum LILC is presented in Chapter **6**. However, the deep analysis was impossible because the etching layer of the aluiminum film is quite difficult. Seven deep problems of LILC were confirmed by prototyping. The result of the development of the novel charactristic equations for the decoupling component is presented in Chapter **7**. It is based on the electromagnetism including EMW theory and the transmission line technologies. The accuracy of the development of the calculation equation was confirmed by compering to the measured data of LILC and the commercialized similar component. As the result, the degrading of the transmission coefficient at mounting on PCB was confirmed. The development result of the novel on-board decoupling component is presented in Chapter **8**. The novel on-board low-impedance lossy line (LILL) was developed by applying the developed characteristic equations. The problems of the LILC mentioned in the conclusions of Chapter **6** were were solved by LILL applied the suitable materials and formation. The on-board LILL has the immediate effectivity because it has excellent manufacturability, availability, and the cost performance for the power decoupling component. In addition, it has the effectiveness which stabilizes the performance of the LSI on PCB. The novel solitary electromagnetic wave theory (SEMW theory) is presented in Chapter **9**. The switching device forming SMC cannot generate the continuous wave because it only operates in the switching period. In relatively recent years, the nonlinear undulation theory was presented and has been developed in physics. The narrow water way and the gate were used at the experiment by John Scott Russell who first found the solitary wave. It was considered that the narrow water way corresponds to the transmission line and the gate corresponds to the switching transistor in the SMC. The generation mechanism and behaviour of the SEMW on the transmission line are presented. The conversion between the time domain and the frequency domain is possible by the modified significant frequency (MSF) defined in SEMW theory. As above, the analyses of SMC in accordance with the electromagnetism become very accurate and very efficient by applying SEMW theory. The novel on-chip LILL technology is presented in Chapter **10**. The on-chip LILL became possible by the SEMW theory. When the LILL is located near the on-chip inverter, on-chip LILL only have to handle SEMW and the rise time of the signal will be shortened close to the gate delay of the on-chip inverter. EMC will be improved greatly when the on-chip LILL technologies are applied to all LSI. The novel matched impedance lossy line (MILL) technology is presented in Chapter **11**. The MILL became possible by the SEMW theory. The performance of MILL was confirmed by both measurement and the calculation. MILL is used to the signal line. The characteristic impedance of MILL is designed to be almost equal to it of the signal line. SEMW travels on MILL smoothly with being attenuated but the wave length of SEMW is maintained between input and output. As the result, the rise/fall time and the magnitude of the signal voltage are maintained. The crosstalk and reflection bounce are redused because MILL attenuates SEMW on the signal line. The innovative circuit and system technologies are presented in Chapter **12**. The quasi-stationary state is defined by the electromagnetism. EMI problem does not exist in QSCC. SPICE can be used at ease if all parts of SMC are reconfigured to the group of QSCC, because the action of SEMW can be ignored. The complete QSCC of SMC will be formed by LILL technologies and MILL technologies for the first time in the world. Recently the serial data transmission such as USB is used to the interface between the ITE, however the high-speed parallel data transmission will be actualized because the crosstalk is suppressed. IT which is supported by SMC as well as SMC will be innovated by the QSCC reconfiguration of SMC.

2

Send Orders for Reprints to reprints@benthamscience.net

Switching Mode Circuit Analysis and Design, 2013, 3-14

CHAPTER 1

Overview of Circuit Design Technologies about SMC

Abstract: SMC has the relatively short history about 2 centuries, however it became the leading existence in the worldwide industry, academia. The reasons of the improvement and the future problems for improvement are presented.

Keywords: Vacuum tube, diode, computer, electromechanical relay, transistor, IC, MOSFET, CMOS, LSI, IT, SoC, PCB, BGA, Moore's Law, MPU, DRAM, on-chip inverter, on-chip interconnect, repeater.

SMC

SMC has been improved by the improvement of the switching device and became to the leading part of the electronics and electric equipment including the computer. However, the conventional wiring design technologies and the electric components which were developed for the analog circuit have been used till now.

SMC was applied to the telecommunication equipment first. The businessman Samuel F.B. Morse of the United States developed the electrical telegraph and the Morse code which is simple and highly efficient in 1837. The first permanent transatlantic telegraph cable was completed in 1866. However, transatlantic cable was used only few months caused by its failure and the American Civil War. That time, the electric telephone was invented. The technologies of SMC had been used to the switching equipment of the telephone system also.

The origin of the electron device is the vacuum tube. The vacuum tube consists of the glass tube and two or more sealed electrodes. The diode was developed as the oscillation valve by John Ambrose Fleming in 1904. The triode was invented by Lee De Forest in 1907 and the tetrode was invented by Walter H. Schottky in 1919. After this, the multifunction and multisection tubes, the beam power tubes, and the miniature tubes were invented one after another. The vacuum tube was the critical device of the radio communication and broadcasting, radio, television, radar, audio equipment, and computer. The vacuum tube needed a lot of power for the heater in it. Therefore, the size of these facilities or the equipment was very large.

The typical electric equipment in SMC is the computer. The first general-purpose digital computer is the electronic numerical integrator and calculator (ENIAC) which was distributed by John von Neumann in 1946. ENIAC was formed by 18,000 vacuum tubes and 1,500 electromechanical relays. The adding time was 200μs, multiple times of decimal and ten digits were 2.8ms, and dividing time was 6ms. The power consumption was 450kW and the weight was 50ton. The calculation time was greatly improved by ENIAC though the programming of ENIAC was done by manual setting the switches and the plugs. The stored-program architecture which was presented by von Neumann in 1952 was applied to EDVAC. After this, many types of the computer were commercialized in the world.

The first bipolar point-contact transistor was invented by John Bardeen, William Shockley, and Walter Brattain at Bell labs in 1947. The first transistor-based computer was demonstrated at the University of Manchester in 1955. The first silicon transistor was produced by Texas Instruments in 1954. In 1960s, the computer had been largely replaced by transistor-based machines. These computers were smaller, faster, cheaper, and more reliable and these consumed less electric power. The first integrated circuit (IC) was developed by Texas Instruments in 1958.

The first metal-oxide-semiconductor (MOS) field effect transistor (FET) was developed at Bell Labs in 1960. MOSFET is suitable for the IC by its simple structure and low power consumption. The first 4 bit microprocessor named 4004 was commercialized by Intel in 1971. This was developed serendipitously by Intel and Busicon for a Japanese Busicon calculator. 5years after this, 8080 which is the 8-bit microprocessor was commercialized. The initial specified clock frequency limit was $2MHz$. The 8080 microprocessor used non-saturated enhancement load n-channel MOS gates, demanding extra voltages for the load-gate bias. It was manufactured in a silicon gate process using a minimum feature size of $6\mu m$. A single layer of metal was used to the interconnect the approximately 6,000 transistors in the design, but the higher resistance polysilicon layer, which required higher voltage for some interconnects, was implemented with transistor gates. The die size was approximately $20mm^2$. The digital computer consists of the digital circuits. The digital circuit performs basically by the voltage levels corresponding to a logical 0 or 1. An inverter circuit serves as the basic logic gate to swap between those two voltage levels. The complementary MOS (CMOS) inverter was invented by Frank Wanlass while working at Fairchild Semiconductor in 1967. CMOS inverter consists of the n-channel MOS (NMOS) and the p-channel MOS (PMOS). Two important characteristics of CMOS device are high noise immunity and low static power consumption. Processing speed can also be improved due to the relatively low-resistance compared to NMOS-only or PMOS-only type devices. Therefore, the microprocessor, application specific integrated circuit (ASIC), graphics processing unit (GPU), and other logic devices require silicon-based CMOS technologies. The downscaling of minimum dimensions enables the integration of an increasing number of transistors on a single chip, as described by Moore's Law. The essential functions on such large scale integrated circuit (LSI) are data storage and digital signal processing. Now the information technology (IT) industry is being supported by LSI. IT is being used to almost all electronics and electric equipment. Recently, LSI is also called as the system-on-a-chip (SoC) because the function of LSI has been increased and is reaching to the system scale.

PROCESS INTEGRATION AND STRUCTURES OF PCB

PCB or the printed wire board is used to mechanically fix and electrically connect the electronic components such as IC, LSI, capacitor, inductor, resistor, and the connector using the conductive pathways. PCB has several numbers of the copper layer and the insulator layers. The signal traces are formed as the microstrip line which consists of the strip conductor, the ground plane, and the insulator layer. Each strip conductor on the deferent layers is connected by the vias. PCB is used for the equipment and the modules. The size of PCB for the equipment is relatively large and the typical application is the motherboard (MB) and the graphic board for the personal computer (PC) or the TV receiver. The thickness of the copper layer for this application is approximately 1.4*mils* or 55*μm*. The line/space of the trace is 50*μm* to 1mm. FR-4 or FR-5 is the most common material used for the insulator layer today. The number of the copper layer of PCB of the digital circuit is from four to eight usually. The number of the signal trace of the digital circuit is more than several thousands. The layout and the formation of the traces is designed by using CAD system. The capacity of the wire is the critical problem at the wiring design because more numbers of the copper layer increase the cost of the manufacturing. The characteristic impedance of the signal line is designed to be approximately 50Ω. It is sometimes increased to approximately 80Ω for reducing the power consumption.

The size of PCB for the modules is small and the typical application is interposer for LSI packaging. Recently, the technology of PCB is led by the packaging technology of LSI and ASIC. The fine pitch copper wire bond has been introduced into the industry mainstream. The material is replacing to the copper wire from Au wire for cost saving. The diameter of the wire of LSI and ASIC is blow 18μm. The copper wire bond has been in use for power devices with 50μm diameter wires and low I/O counts of IC. The interposer is PCB which is the densely integrated lead-frame and it is used to the ball grid array (BGA). The build-up interposer or the laminate interposer is used for the FC-BGA and the wire bonding (WB) BGA. The line/space of the trace is 12μm /12μm today and it will be 6μm /6μm or 5μm /5μm in 2020.

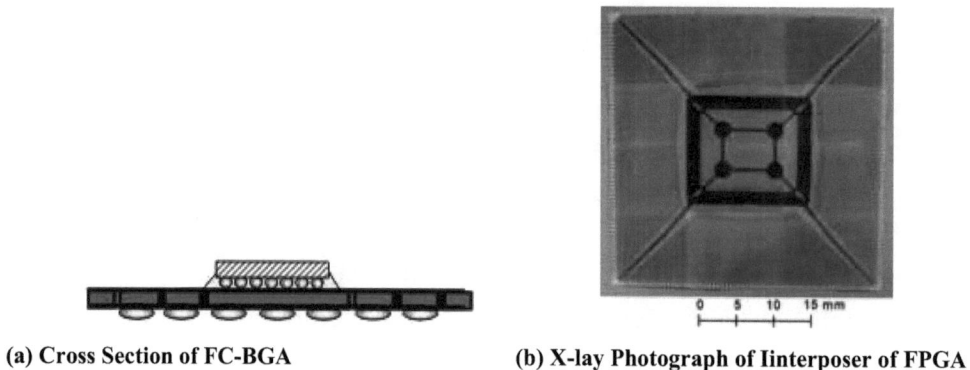

(a) Cross Section of FC-BGA   **(b) X-lay Photograph of Iinterposer of FPGA**

Figure 1: Example of Interposer of FC-BGA.

Fig. **1** shows an example of X-lay photograph of the interposer of the field programmable gate allay (FPGA). The vias interrupt the control of the characteristic impedance at the application of the high-speed digital circuit because the insertion loss occurs at the vias of PCB. The vias are not used on the over gigabit transmission line which is connected between MPU and DRAM for example. DC power is distributed by the power layer and the ground layer which is commonly used for forming the microstrip line. The power traces which have the width as large as possible are formed on the power layer according to several kinds of the supply voltage. The characteristic impedance of the power line is smaller than several ohms. However, its value is not cared by the circuit designer.

Futures Problem

One of the critical problems of the circuit design is considered to be the insertion loss of the transmission line. The formation of the transmission line of PCB can be categorized two groups which consist of 2D transmission line and 5D transmission line. 2D transmission line includes the microstrip line, strip line and the coplanar waveguide. They are formed on a metal layer and the vias are not used. 5D transmission line which includes the microstrip line, strip line and the coplanar waveguide is by the multiple metal layers and the vias which includes the through holes and the blind vias are used. The design of 2D transmission line is not suitable to the densely integration but is relatively easy and the performance at the high-frequency band is integrity is excellent. The improvement of the transmission characteristics of 5D transmission line is necessary to actualize the high-speed and high-densely integration of IT equipment. The improvement of the electrical characteristics of the vias is critical about 5D transmission line. PCB which is embedded the passive component as well as the bear IC chip has been developed.

PROCESS INTEGRATION, DEVICES, AND STRUCTURES (PIDS) OF LSI

According to the international technology roadmap for semiconductor (ITRS), for more than four decades, the semiconductor industry has grown up by the rapid pace of improvement in its products [1].

The principal categories of improvement trends are shown in Fig. **2** with examples of each. All of these improvement trends, sometimes called "scaling" trends, have been enabled by large R&D investments. ITRS reflects the semiconductor industry migration from geometrical scaling to equivalent scaling. Geometrical scaling (such as Moore's Law) has guided targets for the previous 50 years, and will continue in many aspects of chip manufacture. Equivalent scaling targets, such as improving performance through innovative design, software solutions, and innovative processing, will increasingly guide the semiconductor industry in this and the subsequent decade.

TREND	EXAMPLE
Integration Level	Components/Chip, Moore's Law
Cost	Cost per Function
Speed	Microprocessor Clock Rate, GHz
Power	Laptop or Cell Phone Battery Life
Compactness	Small and Light-weight Product
Functionally	Nonvolatile Memory, Imager

Figure 2: Improvement Trends for ICs Enabled by Feature Scaling.

Since its inception in 1992, a basic premise of Roadmap has been that continued scaling of electronics would further reduce the cost per function (historically, ~25–29% per year) and promote market growth for integrated circuits (historically averaging ~17% per year, but maturing to slower growth in more recent history). Thus, Roadmap has been put together in the spirit of a challenge—essentially, "What technical capabilities need to be developed for the industry to stay on Moore's Law and the other trends?" In order to properly represent the continuously evolving facets of the semiconductor industry as it morphs into new and more functional devices in response to the broadening requirements of new customers, 2007 ITRS has addressed the concept of Functional Diversification ("Moore than Moore"). This new definition (MtM) addresses an emerging category of devices that incorporate functionalities that do not necessarily scale in accordance with "Moore's Law", but provide additional value to the end customer in different ways. The default "Time of Introduction" in ITRS is "Year of Production", which is defined as the first two companies reaching production after the presentation of the first conference papers. "Production" means the completion of both process and product qualification. The product qualification means the approval by customers to ship products, which may take one to twelve months to complete after product qualification samples are received by the customer. The trend numbers of ITRS are shown by the half-pitch of the metal 1 of the high-performance MPU/ASIC. The half-pitch of the metal 1 of the dynamic random access memory (DRAM) had been used as the historical indicators of All

IC's scaling. In now a day, the half-pitch of the metal 1 has been used individually now. Here the metal 1 is the fundamental metal layer which fixes the transistor size.

Fig. **3** shows the typical on-chip inverter and its structure.

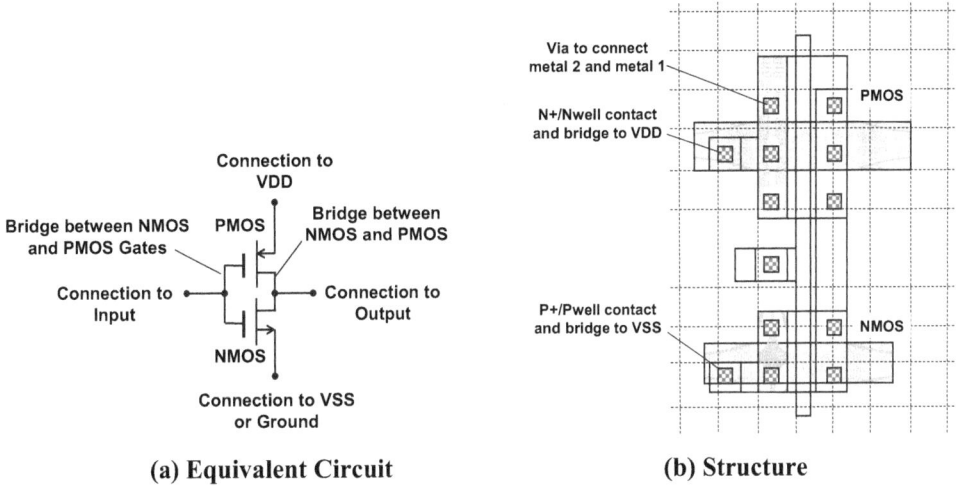

(a) Equivalent Circuit   **(b) Structure**

Figure 3: Structure of On-chip Inverter.

In Fig. **3**, the on-chip inverter is a fundamental circuit of LSI.

The switching speed of MOSFET is limited by the gate delay as well as the gate length which is shown by:

$$\tau = C_g \frac{V_{DD}}{I_{dsat}} \qquad (1)$$

where C_g is the gate capacitance, V_{DD} is the power supply voltage, and I_{dsat} is the saturation current of the drain of MOSFET.

From the equation 1, the gate delay is able to be reduced by reducing the gate capacitance by reducing the gate dielectric thickness, reducing the power supply voltage, and increasing the saturation current of the drain of MOSFET. The thin gate dielectric and large dielectric constant are effective for the speedup of the switching. However they are the difficult challenge because the gate leakage current is increased by them. The reduction of the gate dielectric thickness causes the increase of the gate tunneling current due to band-gap narrowing. The higher permittivity increases the narrowing of the band-gap.

Fig. **4** show the past improvement process of the high-performance MPU/ASIC in accordance with ITRS. Fig. **4a** shows the length of the half-pitch of the metal 1. Fig. **4b** shows the gate delay of PMOSFET. Fig. **4c** shows the drain saturation current ($I_{d,sat}$) of PMOSFET. Fig. **4d** shows the on-chip clock frequency. As shown in Fig. **4a**, **4b**, and **4c**, the technologies of the process integration have been improved steadily. However improvement of the on-chip clock frequency is bogging down since 2007 as shown in Fig. **4d**.

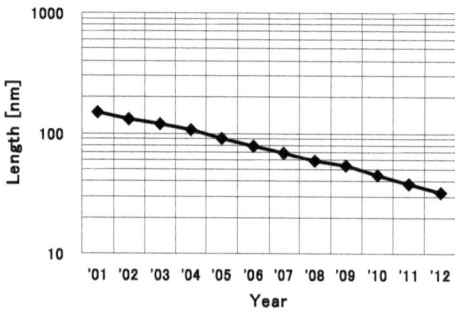

(a) Length of Half-pitch of Metal 1   **(b) Gate Delay of PMOSFET**

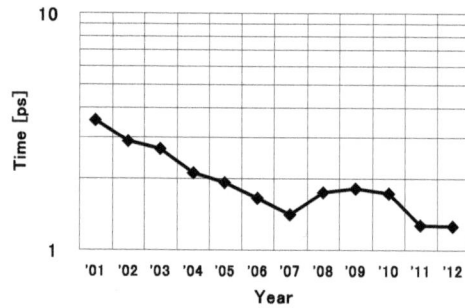

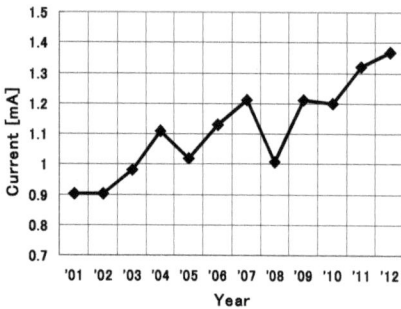

(c) Drain Saturation Current (Id,sat)   **(d) On-chip Clock Frequency**

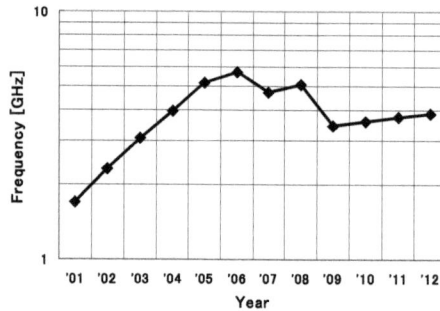

Figure 4: Past Improvement Process of High-performance MPU/ASIC.

According to ITRS 2011 Edition, each on-chip clock frequency in 2016, 2021, and 2026 is estimated to be 4.555GHz, 5.542GHz, and 6.745GHz. Each ring oscillator delay per stage in 2009, 2010, and 2011 was 9.46ps, 9.46ps and 8.04ps.

Each ring oscillator delay per stage in 2012, 2016, 2021, and 2026 is estimated to be 7.62ps, 7.18ps, 2.41ps, and 1.80ps. The delay time is being considered to be improved steadily in the future by supporting the technologies of the process integration. However the on-chip frequency is improved by the technologies of the interconnect design as well as the process integration.

The future problems of PIDS including beyond CMOS are being presented in ITRS.

ON-CHIP INTERCONNECT

Fig. **5** shows the structure example of the on-chip interconnect of LSI which is formed by five copper metallization layers and solder bump used for flip-chip (FC) bonding.

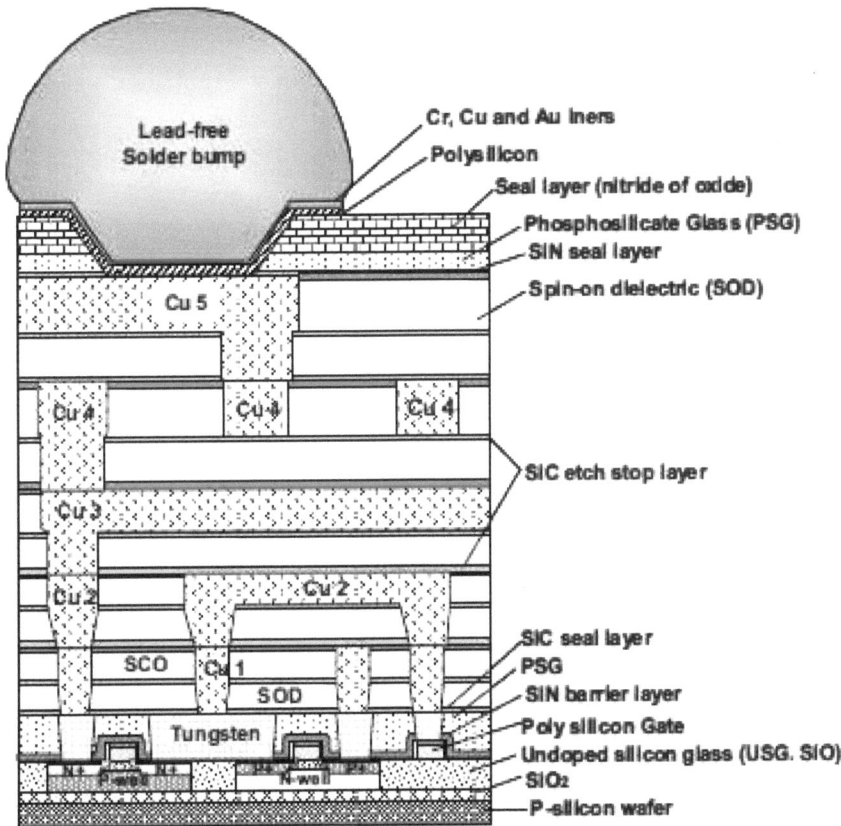

Figure 5: Example Structure of On-chip Interconnect Structure of LSI.

In Fig. **5**, the lead-free solder bump is connected to the interposer which is the multi-layer PCB in the package. Recently, the interconnect of ASIC reaches to fifteen metallization layers which consist of three local layers, four intermediate layers, four global layers, and four power supply wiring layers for example. The global layers are used for the *trans*-chip interconnect on the chip.

Fig. **6** show an example of the two power supply wiring layers.

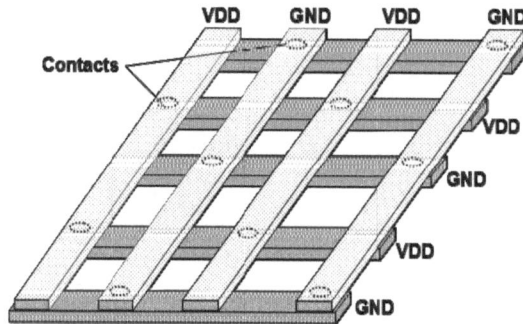

Figure 6: Power Supply Wiring Layers.

An example of the physical and electric parameters of the interconnect is shown in Fig. **7** [1].

Metal Layer (Cu)	Wiring Pitch(nm)		Aspect Ratio		Barrier/cladding thickness(nm)	Delay (ps/mm)	Parasitic Capacitance (pF/cm)
	Min.	Max.	Wire	Via			
Global	154	2,000	2.54	1.5	10	10	1.8-2.1
Intermediate	102		1.8		5.5	1,892	1.9-2.1
Metal 1	102		1.8		5.5	2,100	1.9-2.1
Metal 1 ontacted	90		-		-	-	-

Figure 7: Physical and Electric Parameters of Interconnect.

The low-κ (low dielectric constant) material is used to insulator for reduction of the parasitic capacitance. The idea of the formation of the insulator layer by forming the vacuum void was presented for this purpose. According to 2011 ITRS, the average dielectric constant of inter-level metal insulator are considered to be to 2.82-3.16 in 2012, to be 2.55-3.0 in 2015, to be 2.2-2.46 in 2020, and to be1.7-2.27 in 2026. The copper is used to the wire for the reduction of the resistance instead of aluminum. The on-chip interconnect has been designed based on AC circuit theory and has been applied the lumped element model because the length of the on-chip interconnect are short enough. The resistance and the capacitance are only used usually because the calculation result spreads out when the inductance is used with these parameters. The delay time is calculated by the time constant of the capacitance and the resistance. The repeaters have been often used to shorten the delay time of the global interconnect which connect across the chip.

Fig. **8** shows the model and the equivalent circuit for the SPICE analysis. Fig. **8a** shows the transmission line model and Fig. **8b** shows the CR model.

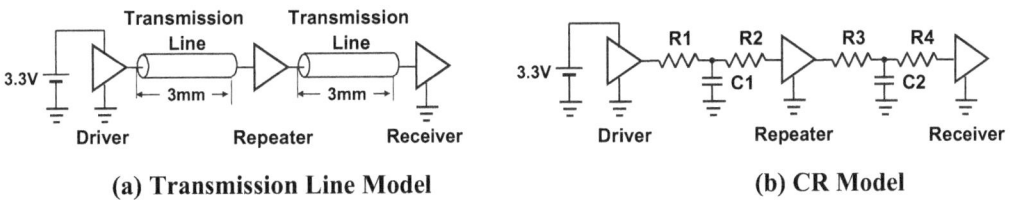

(a) **Transmission Line Model**   (b) **CR Model**

Figure 8: Simulation Model about Performance of Repeater.

In Fig. **8**, the rise time of the driver was $1ps$, the source voltage was set to $3.3V$, the output resistance of the driver and the repeater were 55Ω or 0.55Ω, the gate capacitance of the receiver and the repeater were set to $0.4pF$ or $0.004pF$, respectively. In Fig. **8a**, each simulated characteristic impedance, the loss resistance, and the length z of the transmission line was 86Ω, $0.1\Omega/\mu m$, and $5mm$. In Fig. **8b**, each resistance of R1 to R4 and capacitance of C1 and C2 is 150Ω and $0.182pF$.

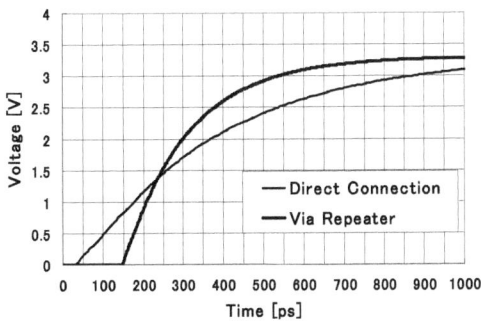

(a) $R_D = 55\Omega$ and $C_R = 0.4pF$ (b) $R_D = 0.55\Omega$ and $C_R = 0.004pF$

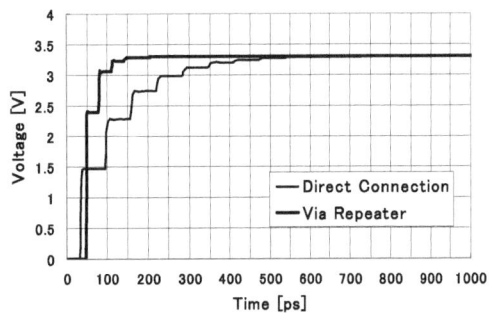

Figure 9: Simulated Voltage shape of Rising Part.

Fig. **9** shows the voltage shape of the receiver. Fig. **9a** shows the simulated voltage shape at the receiver when the gate capacitance at the receiver (C_R) is 0. $4pF$ and the output resistance of the driver (R_D) is 55Ω. Fig. **9b** shows the voltage shape of the rising part at the receiver when the gate capacitance of the receiver is $0.004pF$ and the output resistance of the driver is 0.55Ω, respectively.

The power supply wiring layer is designed as the mesh structure from the point of view of DC circuit. The crosstalk problem has been almost solved by the shortened wire length and the logical techniques. The other problems are caused by EMI. The electromagnetic noise on the power supply wire and the substrate of the chip is the critical problem of LSI. In addition, the electromagnetic noise is considered to be generated by the inductance of the lead-frame. The solution of these problems has been considered to be difficult because the conventional electromagnetism has not been believed unfortunately. The futures problem of the on-chip interconnect are presented in ITRS. For example, the on-chip transmission line interconnect [2, 3] and the optical interconnect [4, 5] were proposed as the solution of the enhancement the performance and the EMI problem.

REFERENCES

[1]   http://www.itrs.net/reports.html
[2]   K. Masu, *et al.*, "On-Chip Transmission Line Interconnect for Si CMOS LSI. IEEE Silicon Monolithic Integrated Circuits in RF Systems, 2006 Topical Meeting, pp. 353- 356, 2006.
[3]   A. Mineyama, *et al.*, "LVDS-type On-Chip Transmission Line Interconnect with Passive Equalizers in 90nm CMOS Process", IEEE ASPDAC 2008 Asia and South Pacific, pp. 970- 98, 2008.
[4]   H. Cho, *et al.*, "Power comparison between high-speed electrical and optical interconnects for interchip communication", IEEE Journal of Lightwave Technology, vol. 22, Issue 9, pp. 2021-2033, 2004.
[5]   Rakheja, S.; Kumar, V., "Comparison of electrical, optical and plasmonic on-chip interconnects based on delay and energy considerations", IEEE 13th ISQED, pp. 732- 739, 2012.

Send Orders for Reprints to reprints@benthamscience.net

CHAPTER 2

Electromagnetic Analysis of PCB

Abstract: The electromagnetic characteristics of many types of the microstrip line and the shielded vias are analyzed by using the method of 3-dimension finite-difference time-domain (FDTD) simulator.

Keywords: Microstrip line, straight microstrip line, 135-degree angle microstrip line, right angle microstrip line, radiation pattern, parallel straight microstrip lines, near-end crosstalk, parallel 135-degree angle microstrip lines, parallel right angle microstrip lines, radiation efficiency, antenna efficiency, electric field (EF) distribution, transmission coefficient, Gaussian pulse, radiation pattern, vias, FR-4, PCB, shielded via, reflection coefficient, vertical transition.

ANALYSIS OF CONVENTIONAL TRANSMISSION LINE

The characteristics of the transmission line on PCB have been analyzed from every aspect [1-6]. The transmission characteristics, electromagnetic radiation, and the crosstalk of many forms of the transmission line were reconfirmed for applying to search the problems of the circuit design and EMC design.

Fig. **1** shows the formation and the simulated characteristics of the Model #1. Fig. **1a** shows the formation. Fig. **1b** shows the EF distribution of the surface. Fig. **1c** shows the transmission coefficient.

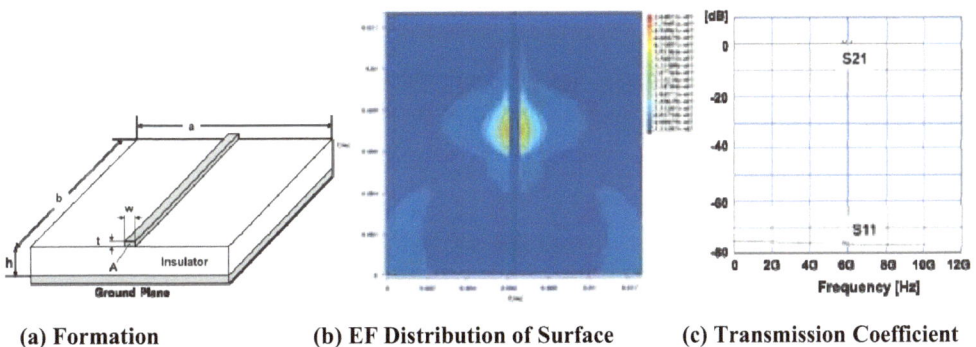

| (a) Formation | (b) EF Distribution of Surface | (c) Transmission Coefficient |

Figure 1: Formation and Simulated Charateritics of Model #1.

In Fig. **1**, the Gaussian pulse was used at the simulation, the model #1 corresponds to the straight microstrip line, and the simulation condition was following; a was 12.7*mm*, b was 12.7 *mm*, h was 635*μm*, t was 63.5*μm*, w was 635*μm*, the dielectric constant of the insulator was 4, and A was the injection port of the Gaussian pulse of which arrival time to the peak is 4×10^{-11} second, pulse width is 1×10^{-11} second, and magnitude is $1V$, respectively. In Fig. **1b**, the integrity of the wave is being maintained on the microstrip line. In Fig. **1c**, transmission coefficient S_{21} is 0dB to up to $10GHz$, and the reflection coefficient is small enough. From above, the straight microstrip line is considered to be the excellent transmission line.

Fig. **2** shows the formation and the simulated characteristics of the Model #2. Fig. **2a** shows the formation. Fig. **2b** shows the EF distribution of the surface. Fig. **2c** shows the transmission coefficient.

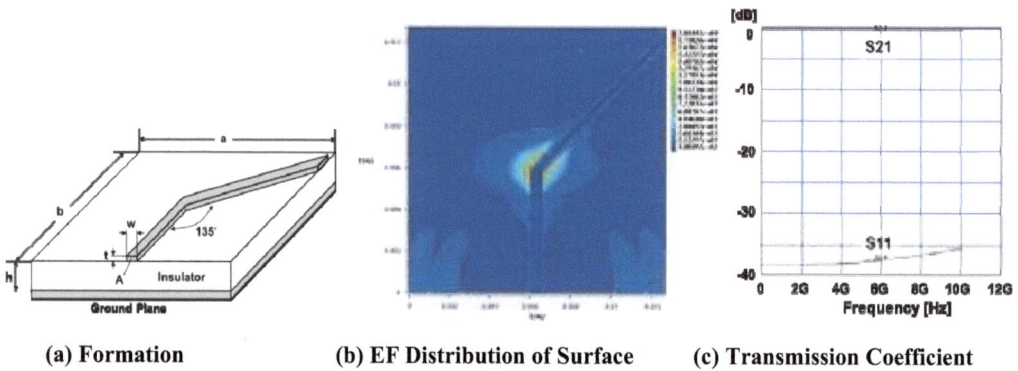

| (a) Formation | (b) EF Distribution of Surface | (c) Transmission Coefficient |

Figure 2: Formation and Simulated Characteritics of Model #2.

In Fig. **2**, the Gaussian pulse line was used at the simulation, the model #2 corresponds to the 135-degree angle microstrip line, and the simulation condition is equal to the model #1. In Fig. **2b**, the integrity of the wave is being almost maintained at the point of the 135-degree angle. In Fig. **2c**, the reflection coefficient is increasing with approaching to $10GHz$ but the magnitude is small enough.

Fig. **3** shows the formation and the simulated characteristics of the Model #3 Fig. **3a** shows the formation. Fig. **3b** shows the EF distribution of the surface. Fig. **3c** shows the transmission coefficient.

In Fig. **3**, the Gaussian pulse line was used at the simulation, the model #3 corresponds to the right angle microstrip line, and the simulation condition is equal to the model #1. In Fig. **3b**, the integrity of the wave is varying at the point of the right angle. In Fig. **3c**, the reflection coefficient of the model #3 is relatively large.

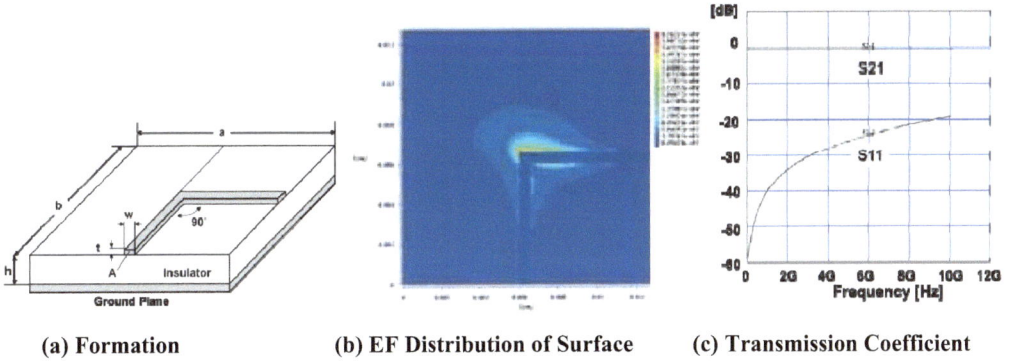

| (a) Formation | (b) EF Distribution of Surface | (c) Transmission Coefficient |

Figure 3: Formation and Simulated Characteritics of Model #3.

Fig. **4** shows the calculated radiation pattern of the surface of the model #1. Fig. **4a** shows it of $100MHz$. Fig. **4b** shows it of $1GHz$. Fig. **4a** shows it of $10GHz$.

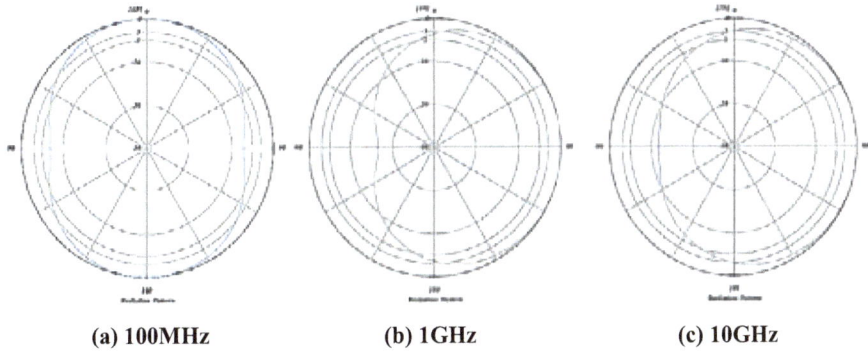

| (a) 100MHz | (b) 1GHz | (c) 10GHz |

Figure 4: Calculated Radiation Pattern of Surface of Model #1.

Each Fig. **5a** and **5b** shows the calculated electric field distribution of the surface on the model #2 at $100MHz$ and $10GHz$. Fig. **5c** shows the calculated electric field distribution of the surface on the model #3 at $10GHz$.

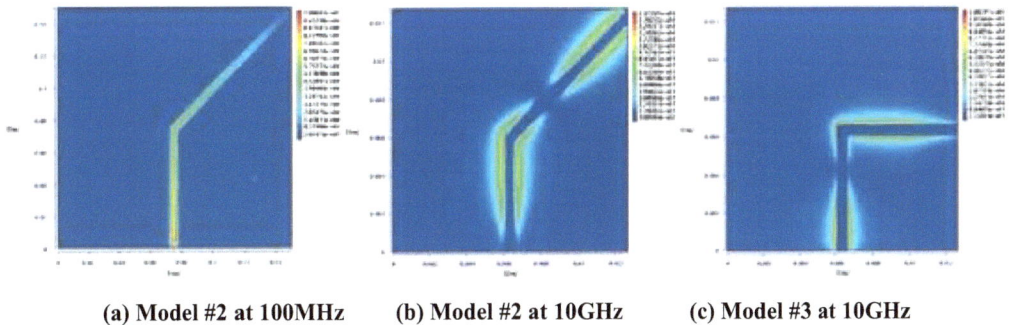

| (a) Model #2 at 100MHz | (b) Model #2 at 10GHz | (c) Model #3 at 10GHz |

Figure 5: Calculated Electric Field Distribution of Surface of Models.

In Fig. **5a**, the side length a and b of the model #2 were modified to 100mm from 12.7mm at 100MHz. In Fig. **5b** and **5c**, the side length a and b of the model #2 and model #3 were modified to 150mm from 12.7mm. The calculated radiation efficiency or the antenna efficiency of the model #2 was 2.19×10^{-7} at 100MHz, 4.85×10^{-4} at 1GHz, and 1.321×10^{-2} at 10GHz, respectively. The calculated radiation efficiency or the antenna efficiency of the model #3 was 3.04×10^{-7} at 100MHz, 1.787×10^{-4} at 1GHz, and 1.885×10^{-2} at 10GHz.

Fig. **6** shows the formation and the simulated characteristics of the Model #4 Fig. **6a** shows the formation. Fig. **6b** shows the EF distribution of the surface. Fig. **6c** shows the transmission coefficient.

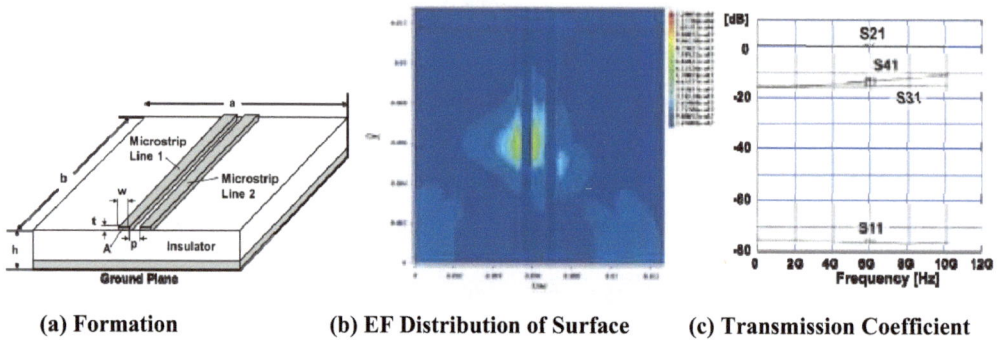

| (a) Formation | (b) EF Distribution of Surface | (c) Transmission Coefficient |

Figure 6: Formation and Simulated Characteritics of Model #4.

In Fig. **6**, the Gaussian pulse line was used at the simulation. In Fig. **6a**, the model #4 consists of parallel straight microstrip lines which are the microstrip line 1 and 2, and the simulation condition was the following; p was 635μm which is equal to w, and the other parameters are equal to them of model #1, respectively. In Fig. **6b**, almost the electric field of the microstrip line 1 is covering the microstrip line 2. In Fig. **6c**, S_{41} which is the far-end crosstalk is increasing to -11dB from -15dB, and S_{31} which is the near-end crosstalk is -15dB to 10GHz from 0Hz.

Fig. **7** shows the formation and the simulated characteristics of the Model #5 Fig. **7a** shows the formation. Fig. **7b** shows the EF distribution of the surface. Fig. **7c** shows the transmission coefficient.

In Fig. **7a**, the model #5 consists of parallel 135-degree angle microstrip lines which are the microstrip line 1 and 2, and the simulation condition is equal to the model #4. In Fig. **7b**, almost the electric field of the microstrip line 1 is covering the microstrip line 2. In Fig. **7c**, S_{31} which is the near-end crosstalk is -15dB but S_{41} which is the far-end crosstalk increasing till -10dB at $10GHz$. However, the crosstalk of the model #5 is smaller than it of the model #4.

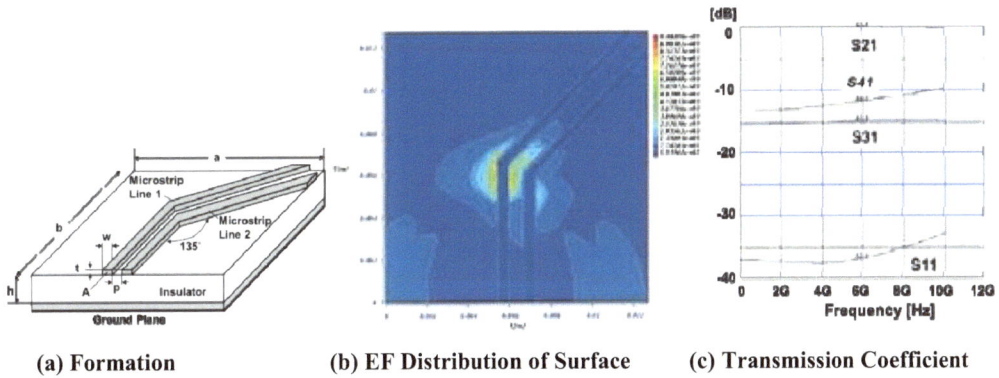

| (a) Formation | (b) EF Distribution of Surface | (c) Transmission Coefficient |

Figure 7: Model #5 of Transmission Line and Simulated Electric Field of Gaussian Pulse.

Fig. **8** shows the formation and the simulated characteristics of the Model #6 Fig. **8a** shows the formation. Fig. **8b** shows the EF distribution of the surface. Fig. **8c** shows the transmission coefficient.

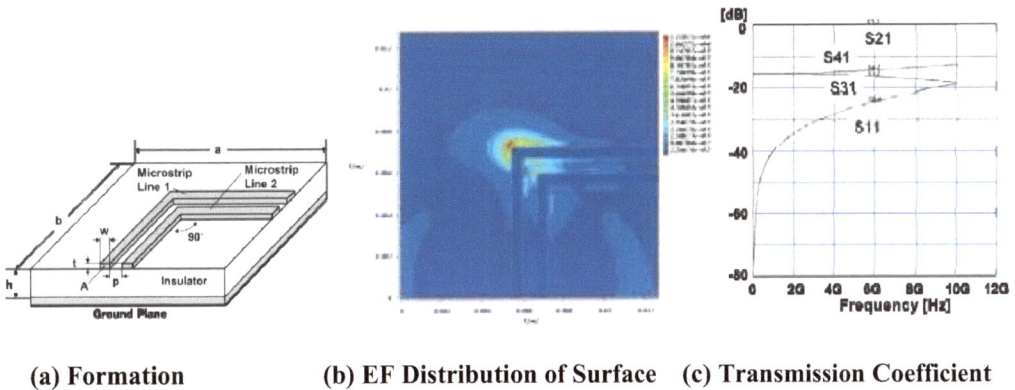

| (a) Formation | (b) EF Distribution of Surface | (c) Transmission Coefficient |

Figure 8: Model #6 of Transmission Line and Simulated Electric Field of Gaussian Pulse.

In Fig. **8a**, the model #6 consists of parallel right angle microstrip lines which are the microstrip line 1 and 2, and the simulation condition is equal to model #4. In Fig. **8b**, almost the electric field of the microstrip line 1 is covering the microstrip line 2. In Fig. **8c**, S_{31} which is the near-end crosstalk is decreasing to -18dB from

-15dB but S_{41} which is the far-end crosstalk increasing to -13dB from -15dB. The crosstalk of the model #6 is smaller than it of the model #4.

Each Fig. **9a**, **9b**, and **9c** shows the calculated EF distribution of the surface of the model #4, #5, and #6 at 10*GHz*.

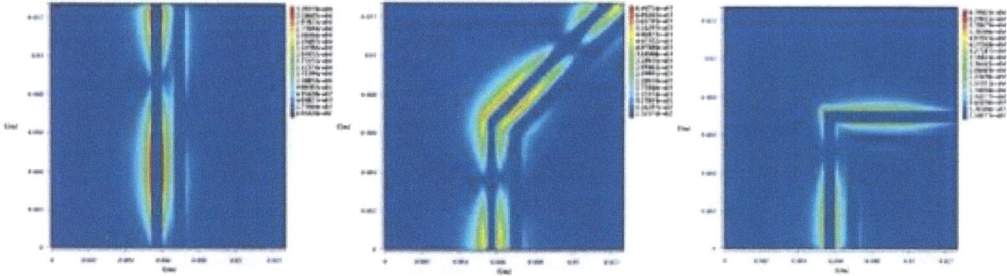

(a) *At 10GHz* of Model #4   (b) At *10GHz* of Model #5   (c) At *10GHz* of Model #6

Figure 9: Calculated Electric Field Distribution of Surface of Models.

Each Fig. **10a**, **10b**, and **10c** shows the calculated radiation pattern of the surface of the model #4, model #5, and model #6 at 10*GHz*. The calculated radiation efficiency or the antenna efficiency of the model #5 was 9.34×10^{-6} at 100*MHz*, 4.56×10^{-4} at 1*GHz*, and 2.01×10^{-2} at 10*GHz*. The calculated radiation efficiency or the antenna efficiency of the model #6 was 3.16×10^{-7} at 100*MHz*, 1.735×10^{-4} at 1*GHz*, and 2.01×10^{-2} at 10*GHz*, respectively.

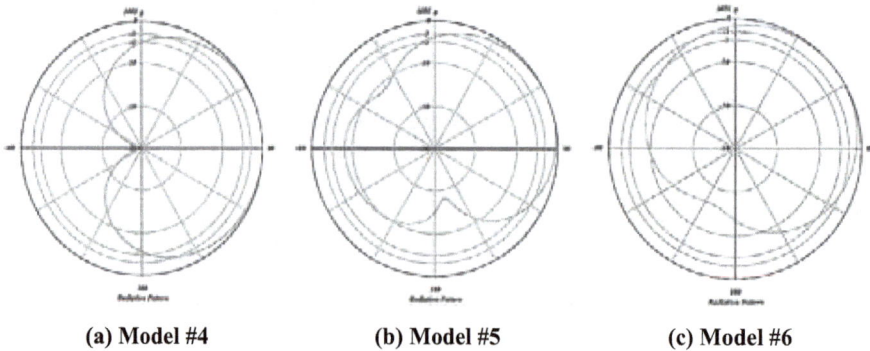

(a) Model #4   (b) Model #5   (c) Model #6

Figure 10: Calculated Radiation Pattern of Surface at 10*GHz*.

Each Fig. **11a** and **11b** shows the model #7 of the transmission lines and the simulated transmission coefficient of this model.

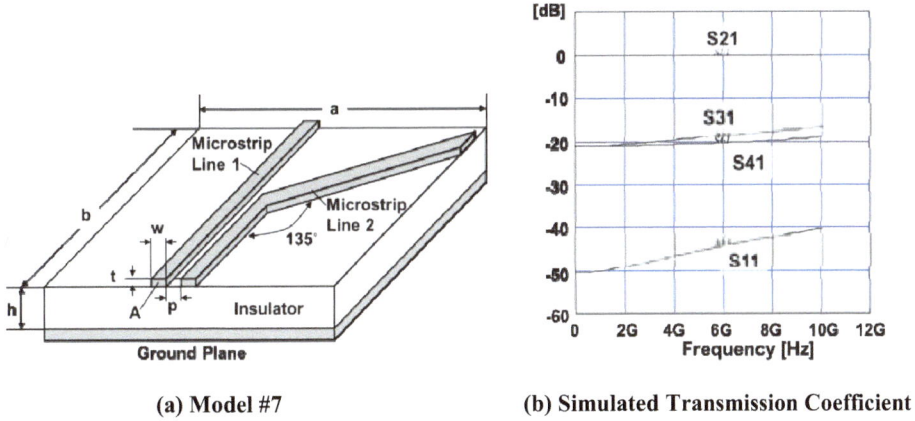

(a) Model #7   (b) Simulated Transmission Coefficient

Figure 11: Model #7 and Simulated Transmission Coefficient of this Model.

In Fig. **11a**, the model #7 consists of parallel microstrip lines which are the straight microstrip line 1 and the 135-degree angle microstrip line 2, and the simulation condition is equal to the model #4. In Fig. **11b**, S_{31} which is the near-end crosstalk is increasing to -16dB from -21dB and S_{41} which is the far-end crosstalk is increasing to -19dB from -21dB. The crosstalk of the model #7 is small enough.

Each Fig. **12a** and **12b** shows the model #8 of the transmission lines and the simulated electric field of it.

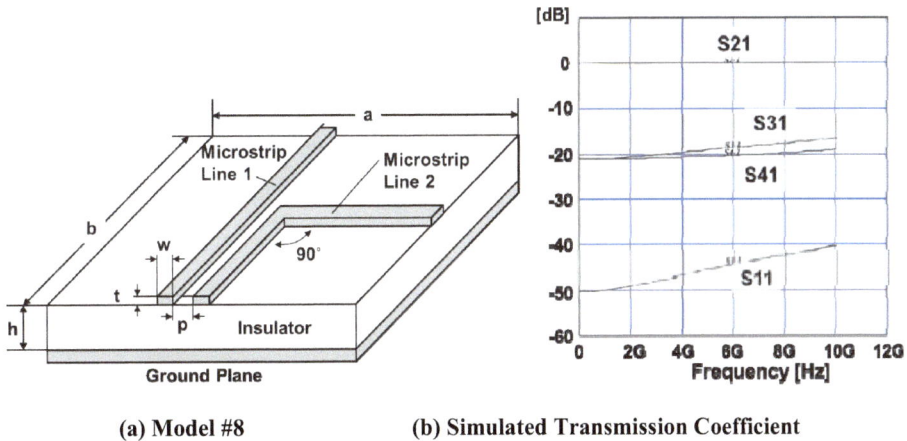

(a) Model #8   (b) Simulated Transmission Coefficient

Figure 12: Model #8 and Simulated Transmission Coefficient of this Model.

In Fig. **12a**, the model #8 consists of parallel microstrip lines which are the straight microstrip line 1 and the right angle microstrip line 2, and the simulation condition is equal to the model #4. In Fig. **12b**, S_{31} which is the near-end crosstalk and S_{41} which is the far-end crosstalk is larger than them of the model #7.

Each Fig. **13a** and **13b** shows the calculated EF distribution of the surface of the model #7 and model #8 at $10GHz$.

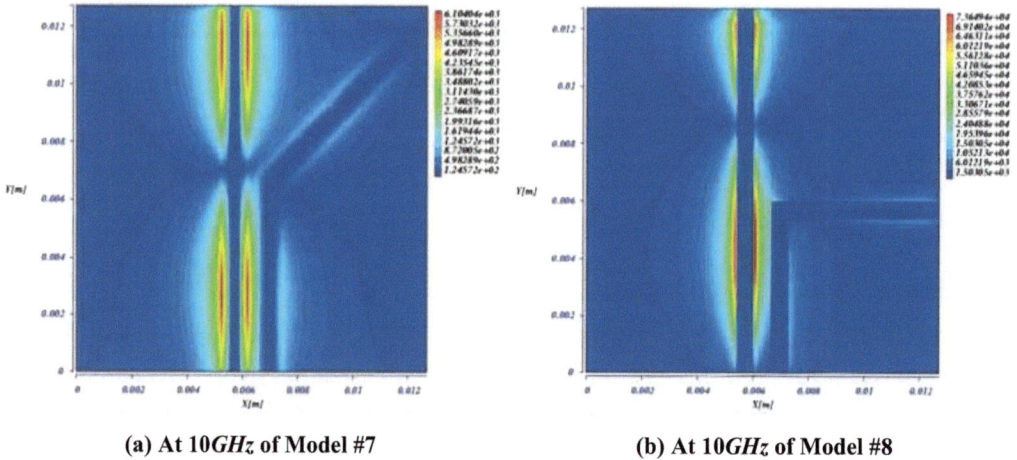

| (a) At $10GHz$ of Model #7 | (b) At $10GHz$ of Model #8 |

Figure 13: Calculated Electric Field Distribution of Surface of Models.

Each Fig. **14a** and **14b** shows the calculated radiation pattern surface of the model #7, and model #8 at $10GHz$. The calculated radiation efficiency or the antenna efficiency of the model #7 was 8.59×10^{-7} at $100MHz$, 1.888×10^{-4} at $1GHz$, and 2.08×10^{-2} at $10GHz$. The calculated radiation efficiency or the antenna efficiency of the model #8 was 4.95×10^{-7} at $100MHz$, 1.940×10^{-4} at $1GHz$, and 2.08×10^{-2} at $10GHz$.

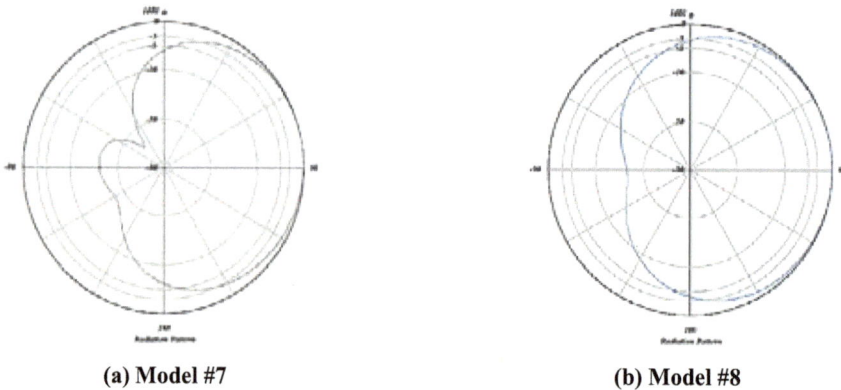

| (a) Model #7 | (b) Model #8 |

Figure 14: Calculated Radiation Pattern of Surface at $10GHz$.

ANALYSIS OF SHIELDED *VIA* [7-9]

The vias on the microstrip line decrease the transmission coefficient.

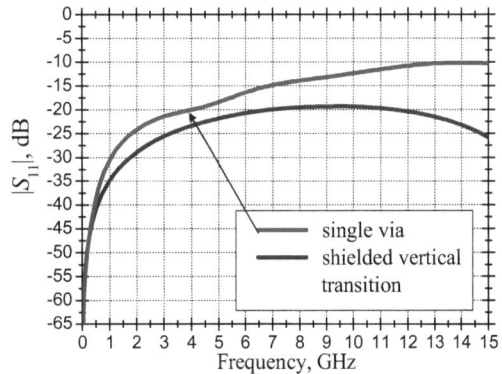

Fig. **15** shows the simulation model of the single *via* and the shielded via.

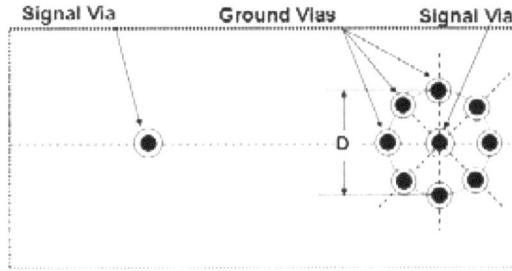

Figure 15: Simulation Model of Single Via and Shielded Via.

In Fig. **15**, the diameter of the drilled hole was 0.65*mm*, the diameter of the pad was 1.65*mm*, the distance between the nearest signal *via* and ground *via* was 1.63*mm,* the thickness of PCB was 2.5*mm,* and the relative permittivity and loss tangent of FR-4 filling the PCB were 4.2 and 0.023, respectively.

Fig. **16** shows the simulated electrical performance of the single *via* and the shielded vertical transition with the square arrangement of the ground vias in a 12-conductor layer PCB. In Fig. **16** the simulated electrical performance was measured by using the 50Ω coaxial cables.

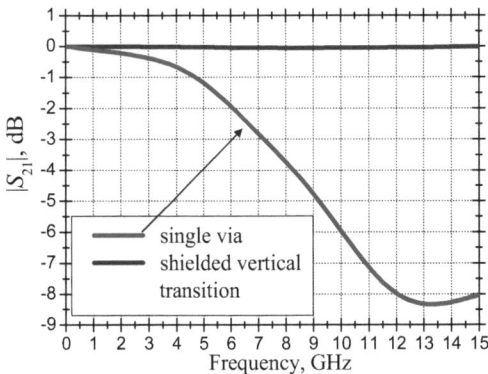

(a) Transmission Coefficient S_{21}   **(b) Reflection Coefficient S_{11}**

Figure 16: Simulated Electrical Performance of Vias.

Fig. **17** shows the improved shielded vias. Fig. **17a** shows the top view and Fig. **17b** shows the cross section view. $D = 2.52mm$, $d_{a,1} = 0.5mm$, $d_{a,2} = 0.5mm$, and $d = 0.65mm$. In Fig. **17a**, the seven black circles mean the ground vias, and the seven white circles mean the air holes, respectively. The transmission coefficient was simulated between Port 1 and Port 2 which is a terminal of the 50Ω strip line formed in PCB which consists of 8-conductor layer PCB.

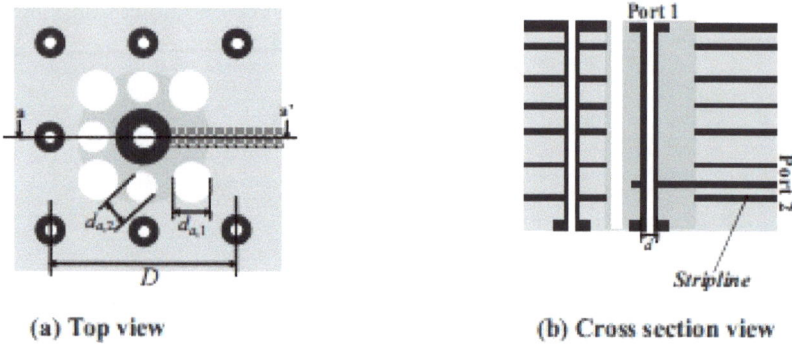

| (a) Top view | (b) Cross section view |

Figure 17: Improved Shielded Vias.

Fig. **18** shows the simulated characteristics of the improved shielded vias. Each Fig. **18a** and **18b** shows S_{21} and S_{11}. S_{11} and S_{21} of the shielded vias with air holes are considered to be small enough.

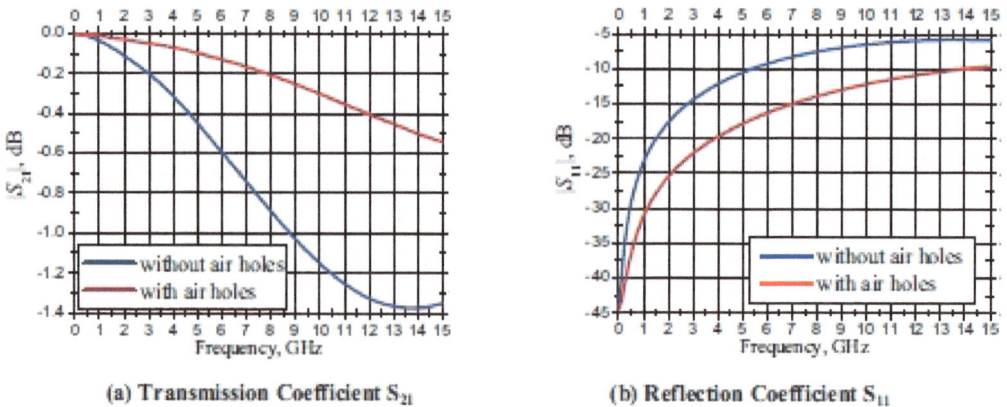

| (a) Transmission Coefficient S_{21} | (b) Reflection Coefficient S_{11} |

Figure 18: Simulated Charectristics of Improved Shielded Vias.

Fig. **19** shows the experimental pattern of the compact vias with four air holes in the multilayer PCB.

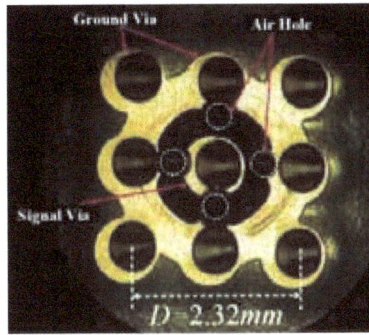

Figure 19: Experimental Pattern of Compact Vias with Four Air Holes.

In Fig. **19**, the diameter of the air hole was 0.3mm.

Fig. **20** shows the simulation and measurement result of the prototyped Vias shown in Fig. **19**.

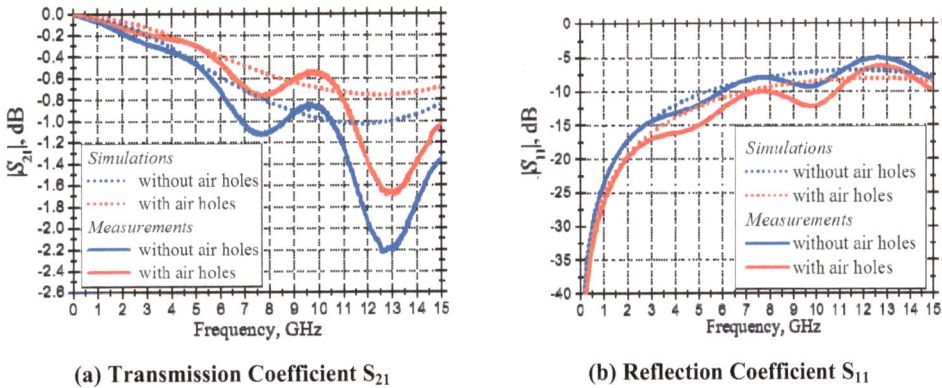

(a) **Transmission Coefficient S$_{21}$**   (b) **Reflection Coefficient S$_{11}$**

Figure 20: Simulated and Measured Charectristics of Improved Shielded Vias.

In Fig. **20**, S_{21} and S_{11} were improved at a frequency higher than 6 GHz approximately, however was little improved at the giga-hertz frequency range.

As the result, the shielded vias or the vertical transitions formed by the signal *via* and the ground vias will provide high-performance signal line on a high-speed multilayer PCB

REFERENCES

[1]   B.C. Wadell, "Transmission Line Design Handbook", Artech House Antennas and Propagation Library, 1991.

[2]   F. Gardiol, "Microstrip Circuits", Wiley Series in Microwave and Optical Engineering, 1994.

[3]   W.T. Joines, W.D. Palmer, J.T. Bernhard, "Microwave Transmission Line Circuits", Artech House Microwave Library, 2012.

[4]   H.-Y. Yee, "Printed Circuit Transmission-Line Characteristic Impedance by Transverse Modal Analysis" IEEE Transactions on Microwave Theory and Techniques, Volume: 34, Issue: 11, pp. 1157-1163, 1986.

[5]   Colpitts, B.G., "Teaching transmission lines: a project of measurement and simulation", IEEE Transactions on Education, Issue: 3, pp. 245-252, 2002.

[6]   S.-W. Jung, Dept. of Electr. Eng., Yeungnam Univ., Gyeongsan-si, Ki-Chai Kim, "Insertion loss of unbalanced transmission line crossing a rectangular aperture in an infinite backplane", IEEE Asia-Pacific Symposium on Electromagnetic Compatibility and 19th International Zurich Symposium on Electromagnetic Compatibility, pp. 494-497, 2008.

[7]   T. Kushta, K. Narita, T. Kaneko, T. Saeki, H. Tohya, "Resonance stub effect in a transition from a through *via* hole to a stripline in multilayer PCB's," IEEE Microwave and Wireless Components Letters, vol. 13, no. 5, pp. 169-171, May 2003.

[8]   K. Narita, T. Kushta, T. Saeki, H. Tohya, "Extracting propagation constants for high-speed digital data transmission on printed circuit boards," in Proc. Int. Conf. on Electronics Packaging, Tokyo, Japan, April 16-18, 2003, pp. 276-281.

[9]   T. Kushta, K. Narita, H. Tohya, "Effect of ground vias on performance of interconnections embedded in multilayer PCB's," in Proc. Int. Conf. on Electronics Packaging, 2003 ICEP, Tokyo, Japan, April 16-18, 2003, pp. 503-508.

Send Orders for Reprints to reprints@benthamscience.net
Switching Mode Circuit Analysis and Design, 2013, 27-38

CHAPTER 3

Analysis of Performance of Decoupling Capacitor

Abstract: Many capacitors are used to the PDN. The conventional measurement method to get the terminal impedance of the capacitor has the problem from the point of view of the electromagnetism.

Keywords: Decoupling capacitor, PDN, parallel connected capacitors, ground plane, chip tantalum capacitor, short-circuit board, transmission coefficient, characteristic impedance, low-pass filter, voltage droop, power supply noise, network analyzer, scattering matrix, 4-terminal circuit, distributed element model, lumped element model, 2-terminal component, propagation constant, reflection loss, PDN.

DECOUPLING PERFORMANCE OF CAPACITOR

The interest to the decoupling circuit of PDN is different between the circuit design and the EMC design. The EMC engineers focus on the best way to increase the decoupling performance of PDN [1-10]. The parallel connected capacitors have been used as the decoupling capacitor on PDN of PCB.

Fig. **1** shows the measured transmission coefficient S_{21} of the transmission line with the parallel connected capacitors.

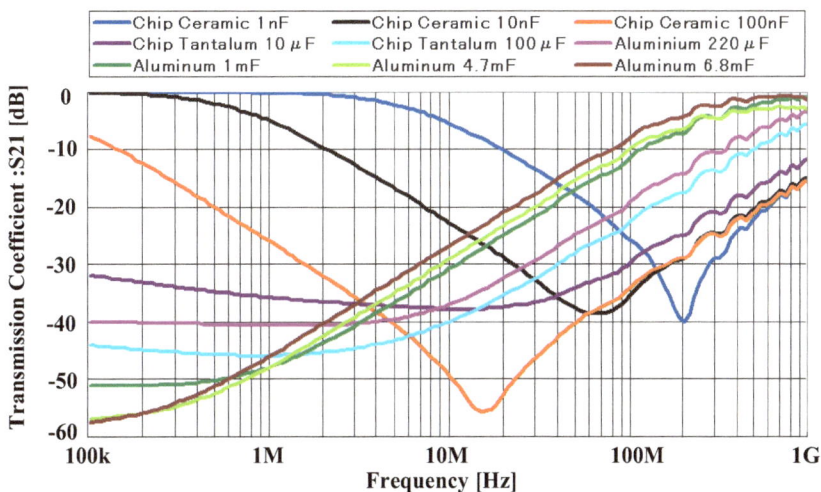

Figure 1: Measured S_{21} of Transmission Line with Parallel Connected Capacitors.

Hirokazu Tohya

In Fig. **1**, the characteristic impedance of the transmission line for measurement was 50Ω. The measured value of S_{21} will increase toward zero when the characteristic impedance of the transmission line for measurement decreases, and the measured value of S_{21} will reduce when the characteristic impedance of the transmission line for measurement increases.

Fig. **2** shows the jigs of the capacitor which were used for measurement with the network analyzer. In Fig. **2**, the capacitors are connected to the planer line which is formed on a surface. The best characteristics have been considered to be able to get by the jig of this formation.

(a) Aluminum Capacitor   (b) Chip Tantalum Capacitor   (c) Chip Ceramic Capacitor

Figure 2: Measurement Jigs of Capacitor.

Fig. **3** show the test board for the measurement of the decoupling performance of the chip tantalum capacitors of $10\mu F$. The chip tantalum capacitor was connected between the strip conductor of the microstrip line and the ground trace which is connected to the ground plane of the back face of the test board by many vias, the characteristic impedance of the microstrip line was 50Ω, and the edges of the microstrip line are connected to the SMA connectors for measuring by the network analyzer. The other similar test board was also prototyped for the capacitance of $100\mu F$.

Figure 3: Test Board for Chip Tantalum Capacitor of 10μF.

Fig. **4** show the typical specification of the chip tantalum capacitor as DUT. These capacitors had been commercialized by NEC.

Name	Size (mm)	Capacitance (µF)	Work Voltage (V)
PSLA	3.2×1.60×1.6	10	6.3
MSVA	3.2×1.60×1.6	10	10
MSVB3	3.5×2.80×1.2	10	10
SVS	2.0×1.25×1.2	10	6.3
PSLD	7.3×4.30×2.8	100	10
MSVD	7.3×4.30×2.8	100	16
SVFD	7.3×4.30×2.8	100	25

Figure 4: Typical Specification of DUT.

Fig. **5** show the short-circuit boards. The short circuit was formed on the space between the strip conductor and the ground pad for connecting DUT.

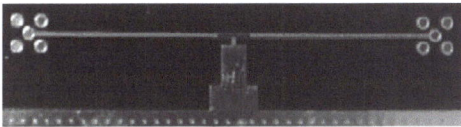

(a) Board #1 for Capacitance of 10µF   (b) Board #2 for Capacitance of 100µF

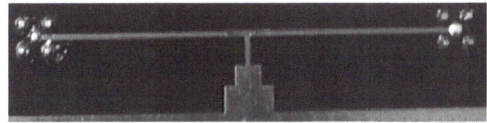

Figure 5: Short-circuit Board.

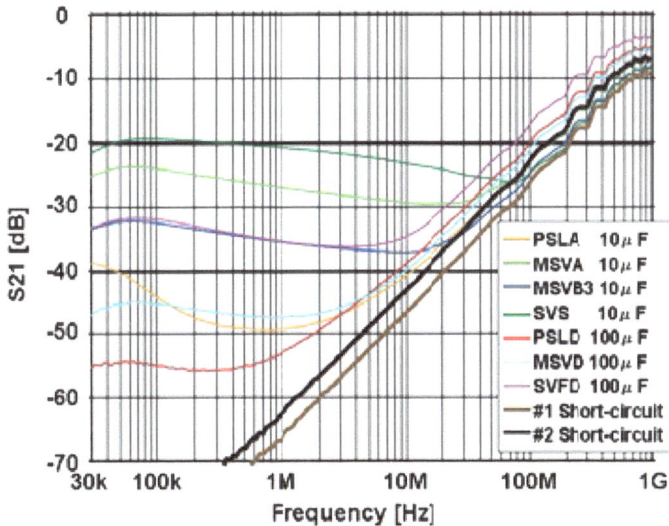

Figure 6: Measured S_{21} of DUT and Short-circuit Board.

Fig. **6** shows the measured S_{21} of DUT and it of the short-circuit boards. Each S_{21} curve of DUT is approaching to it of the short-circuit trace and it is never exceeding the S_{21} curve of each short-circuit board, and each S_{21} is not depending on its capacitance. Therefore, it is supposed that S_{21} of the chip tantalum capacitor depends on the physical formation of the capacitor and the jig at the frequency range of Fig. **3**.

From above, it was clarified that S_{21} of the capacitor depends on the application and the formation of the power conductor of PCB.

Fig. **7** shows the test board for chip tantalum capacitor inserted to the strip conductor.

Figure 7: Test Board for Chip Tantalum Capacitor Inserted to Strip Conductor.

Fig. **8** shows the measured S_{21} of DUT inserted to the strip conductor. In Fig. **8**, the insertion loss of SVS and MSVA is relatively large. However, the insertion loss of all DUT is not large at the actual application. On the other hand, the decoupling performance of S_{21} of SVS and MSVA is relatively small. From this fact, it was considered that this difference of the insertion loss at the use of the serial connection also does not depend on the capacitance but depends on the formation method of the electrode.

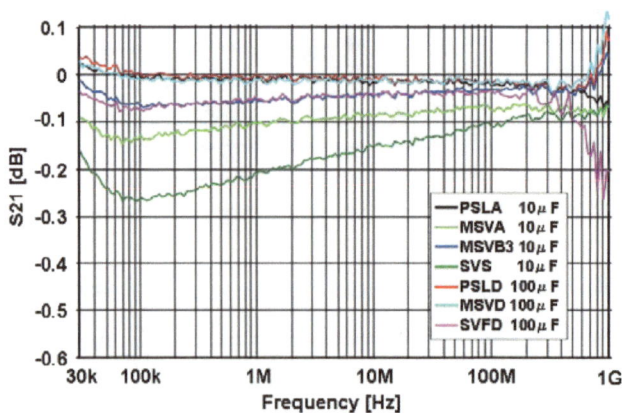

Figure 8: Measured S_{21} of DUT Inserted to Strip Conductor.

Many amount and many kinds of the capacitor have been used to PDN of PCB in ITE. However, decoupling performance of it is not enough to adapt to the EMC regulation without other EMI suppression method. Conventionally, PDN has been formed by the ground plane and the power plane to minimize the terminal impedance at the power terminal of LSI. However, this way is seems to be not suitable to the low-pass filter but the combination of slender power trace and the capacitor will be suitable to the low-pass filter.

PERFORMANCE OF MINIMIZING IMPEDANCE

One of the critical problems at the SMC design of PCB is minimizing the impedance of the PDN to stabilize the performance of LSI. The semiconductor circuit engineers have been required to the PCB design engineers to reduce the voltage droop on PDN from the point of view of the stabilizing of LSI [10]. For this objective, the great numbers and kinds of capacitors are used to PDN as the charm for the low-impedance. That means that the impedance of the PCB depends on the capacitors. In actual, about a thousand capacitors are used to PDN of the original MB of the personal computer and many capacitors are used on the interposer board.

Fig. **9** shows the image of the voltage droop at the LSI on PCB. In Fig. **9**, the 1st droop is considered to be formed by the reactance of the lead-frame in the package and the on-chip capacitor and the duration is of a few nanosecond, and it may severely impact the performance of LSI, the 2nd and 5th droop are considered to be able to be reduced traditionally through improving PCB and package routing and increasing the amount of high-quality decoupling capacitor. In contrast, the 1^{St} droop is considered to be not able to be easily remedied through external measures. In particular, keeping the high-frequency impedance very low requires a significant amount of high-quality decoupling capacitors which are electrically very close to the switching circuits. The power supply noise is proportional to the transient *di/dt* current which has been continued to increase with improving LSI technology. On the other hand, from the point of view of EMC, the large attenuation of the decoupling circuit which consists of the capacitor has been required on PCB.

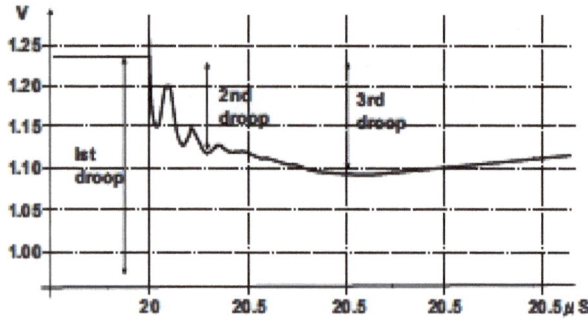

Figure 9: Image of Voltage Droop at LSI on PCB.

Impedance Characteristics of Decoupling Capacitor

The capacitor was invented in 1785 by the Dutch physicist Pieter van Musschenbroek, which was 39 years before the presentation of the coulomb law. The present formation which consists of two conductors separated by the insulator is same to the invented capacitor. The most capacitor is being used in the electric components to the electric and electronics equipment. Almost capacitors in the electric and electronics equipment are used to PDN in parallel as the decoupling capacitor. The decoupling circuit is a kind of low pass filter.

Fig. 10 shows the conventional setup for measuring the admittance Y_t of DUT.

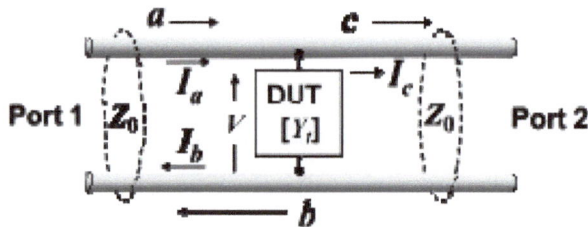

Figure 10: Conventional Setup for Measuring Y_t of DUT.

In Fig. **10**, the port 1 and the port 2 is connected to the network analyzer or the impedance analyzer by the transmission line, the admittance Y_t which is able to consider to be the admittance of the capacitor is placed at the middle of the transmission line, I_a is the injected current, I_b is return current, I_c is passing current, and Z_0 is the characteristic impedance of the transmission line.

Fig. **10** is formed by fusing the idea of the scattering matrix and it of the 4-terminal circuit of the AC circuit theory. Therefore, EMW is replaced to the charge current I_a, I_b and I_c. This replacement does not depend on the electromagnetism but this is not discussed here.

From Fig. **10**, the voltage V, and each current are:

$$V = (a + b)\sqrt{Z_0} = (c + 0)\sqrt{Z_0} \tag{1}$$

$$I_a = (a + b)/\sqrt{Z_0} \tag{2}$$

$$I_c = (c - 0)/\sqrt{Z_0} \tag{3}$$

$$I_a - I_c = Y_i V \tag{4}$$

From the equation 1 to 4,

$$b = \frac{-Y_t}{2Y_0 + Y_t} a \tag{5}$$

$$c = \frac{2Y_0}{2Y_0 + Y_t} a \tag{6}$$

where $Y_0 = 1/Z_0$, and $Y_t = 1/Z_t$.

From the equation 6, transmission coefficient S_{21} is:

$$S_{21} = \frac{2Y_0}{2Y_0 + Y_t} = \frac{2Z_1}{2Z_1 + Z_0} \tag{7}$$

From the equation 7

$$Z_t = \frac{S_{21}}{2(1 - S_{21})} Z_0 \tag{8}$$

When $S_{21} \ll 1$, and $Z_0 = 50\Omega$,

$$Z_t = 25 S_{21} \tag{9}$$

From the equation 8 and 9, the impedance characteristics of the decoupling capacitor can be got from S_{21} of the circuit shown in Fig. **10** by measuring with the network analyzer.

Fig. **11** shows the calculated impedance characteristics of the capacitors, which were transformed by the equation 8 from measured S_{21} shown in Fig. **1**. The calculated impedance was the value when the capacitors are connected to the transmission line of 50Ω, therefore the impedance value should approach to 50Ω at the higher frequency than it of the minimum impedance. However the impedance value at the higher frequency is increasing to more than 50Ω.

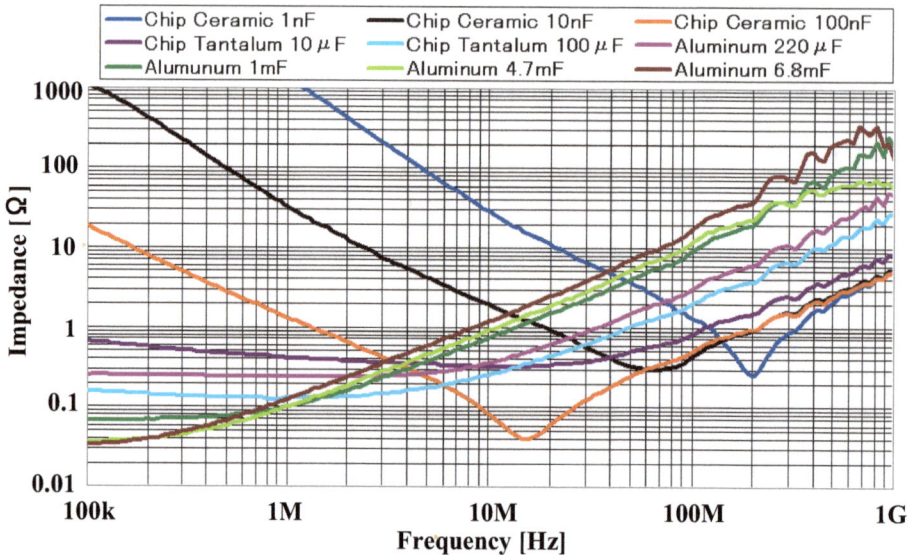

Figure 11: Calculated Impedance of Capacitors from Measured S_{21}.

Relationship of S_{21} and Terminal Impedance of Decoupling Capacitor

The measured S_{21} shown in Fig. **1** is reliable enough. Therefore the fault must exist in the above mentioned equation from 1 to 9. The terminal impedance at the higher frequency than it of the minimum impedance depends on the characteristic impedance of the connected transmission line. The characteristic impedance is defined by the distributed element model circuit. However, the impedance in the equation from 1 to 9 is defined by the lumped element model circuit which cannot handle the electromagnetic wave and the transmission line. The capacitor is one of the 2-terminal components. Therefore, the propagation constant and the reflection coefficient of the decoupling capacitor should be zero. However, only the reflection loss will occur when the structure part of the capacitor disturbs the traveling EMW in the insulator of the transmission line.

Fig. **12** shows the calculated characterisic of the capacitors at the same condition to Fig. **11**.

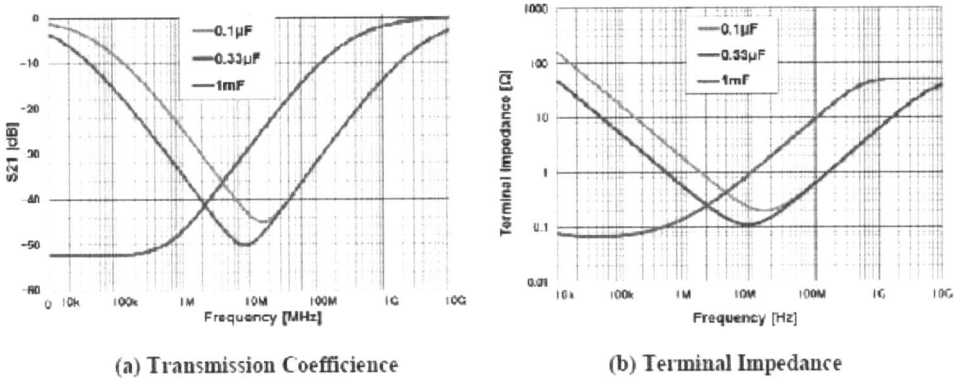

(a) Transmission Coefficience   (b) Terminal Impedance

Figure 12: Calculated Characterisic of Capacitors at Same Condition to Fig. **11**.

In Fig. **12b**, the terminal impedance approaches to 50Ω, and the terminal impedance will approach to 25Ω at the lower frequency than 10kHz when both edges of the transmission line are terminated by 50Ω. These value cannot be got from the equation 7. The calculation condition was as follows; each one capacitor was connected to the center position of the transmission line of $2m$ length, the characteristic impedance of the transmission line was 50Ω.

Fig. **13** shows the calculated characteristics of the capacitors on the conventional PDN which consists of the power plane and the ground plane of PCB, which is the normal form of PDN of PCB.

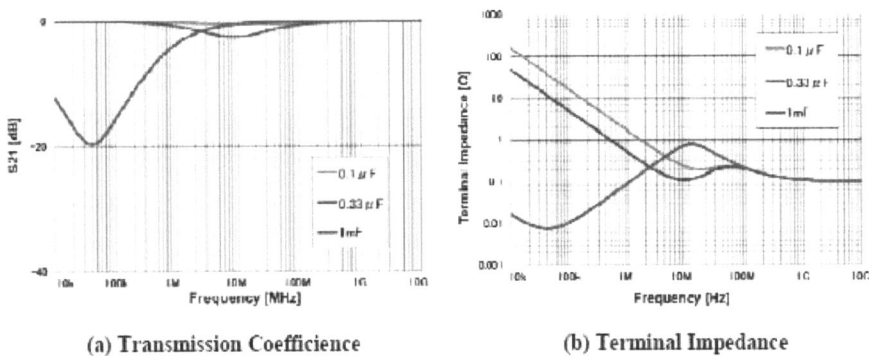

(a) Transmission Coefficience   (b) Terminal Impedance

Figure 13: Calculated Characterisic of Capacitor on Conventional PDN.

In Fig. **13**, S_{21} at a frequency higher than 1*MHz* is little value which means the decoupling performance is so poor, however the terminal impedance at a frequency higher than 1*MHz* is small enough for keeping the performance of LSI, and the calcuration condition was as follows; S_{21} of the capacitors was got from the Fig. **1**, the each one capacitor was used on the center of the plate transmission line consists of the plane of 200*mm*×200*mm*, the ground plane, and the insulator layer of 0.1075*mm* thickness and 4.35 dielectric constant.

Fig. **14** shows the calculated characterisic of the capacitors on the modified PDN which consists of the slander power trace and the ground plane.

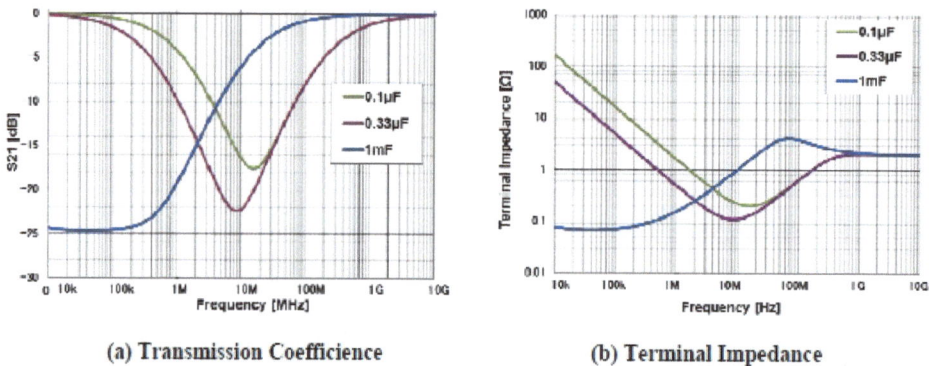

(a) Transmission Coefficience   (b) Terminal Impedance

Figure 14: Calculated Characterisic of Capacitor on Modified PDN.

In Fig. **14**, S_{21} is smaller than it shown in Fig. **13a**, the terminal impedance is larger than it shown Fig. **13b** but this value will be almost enough for keeping the performance of LSI, and the calcuration condition was as follows; S_{21} of the capacitors was got from the Fig. **1**, the each one capacitor was used on the center of each slender power trace of 10*mm*×200 *mm*, the ground plane, and the insulator layer of 0.1075*mm* thickness and 4.35 dielectric constant.

Fig. **15** shows another calculated characterisic of the capacitor on one capacitor on the modified PDN which consists of the branch power trace, stem power trace, and the ground plane.

In Fig. **15a**, S_{21} is smaller than it shown in Fig. **14a**, and the difference of S_{21} between Fig. **15a** and Fig. **13a** at a frequency higher than 10*MHz* is almost 17 dB which value is similar to the improved value of the type-A improved MB of the second workstation. In Fig. **15b**, the terminal impedance is larger than it shown Fig. **13b** and this value will not be enough for keeping the performance of LSI, and the calcuration condition was as follows; S_{21} of the capacitors was got from the Fig. **1**, the each one capacitor was used on the center of the branch power trace of $1mm \times 10mm$ which is connected to the stem power trace of $200mm$ width, the ground plane, and the insulator layer of $0.1075mm$ thickness and 4.35 dielectric constant.

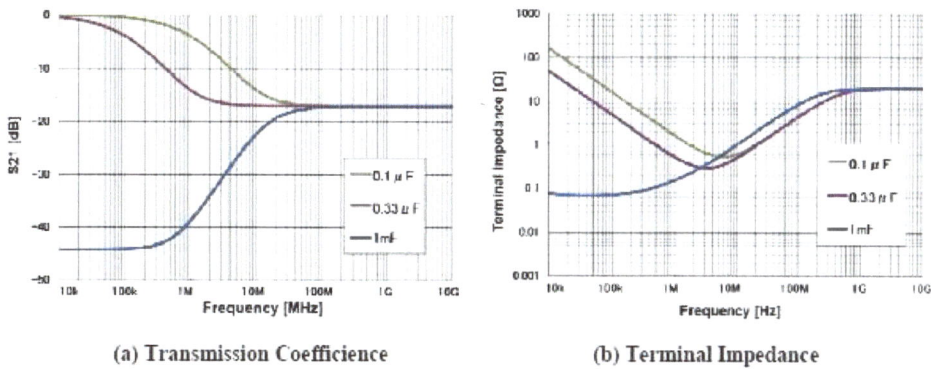

(a) Transmission Coefficence   (b) Terminal Impedance

Figure 15: Calculated Characterisic of Capacitor on Improved MB of 2nd Workstation.

From above, there are no solution when is the capacitor is used to the decoupling circuit of PDN. In addition, there are fewer effects by using many capacitors because the principle of superposition is not valid on the distributed element model circuit but only on the lumped element model circuit.

REFERENCES

[1]   G. Venkataramanan, "Characterization of Capacitors for Power Circuit Decoupling Applications" IEEE Industry Applications Conference, 1998. Thirty-Third IAS Annual Meeting. Vol. 2, pp. 1142-1148, 1998.

[2]   C.B. O'Sullivan, L.D. Smith, D.W. Forehand, "Developing a Decoupling Methodology with SPICE for Multilayer Printed Circuit Boards", 1998 IEEE International Symposium on Electromagnetic Compatibility, vol.2 pp. 652-655, 1998.

[3]   Y.-J. Kim, H.-S. Yoon, S. Lee, G. Moon, J. Kim, J.-K. Wee, "An Efficient Path-Based Equivalent Circuit Model for Design, Synthesis, and Optimization of Power Distribution Networks in Multilayer Printed Circuit Boards, IEEE TRANSACTIONS ON ADVANCED PACKAGING, VOL. 27, NO. 1,pp. 97-106, 2004.

[4]   L.D. Smith, R.E. Anderson, D.W. Forehand, T.J. Pelc, T. Roy, "Power Distribution System Design Methodology and Capacitor Selection for Modern CMOS Technology", IEEE TRANSACTIONS ON ADVANCED PACKAGING, Vol. 22, NO. 3, pp. 284-290, 1999

[5]   J. Fan, J.L. Knighten, A. Orlandi, N.W. Smith, J.L. Drewniak, "Quantifying Decoupling Capacitor Location", 2000 IEEE International Symposium on Electromagnetic Compatibility, vol.2 pp. 761-766, 2000.

[6]   T.M. Zeeff, T.H. Hubing, Orlandi, N.W. Smith, J.L. Drewniak, "Quantifying Decoupling Capacitor Location", 2000 IEEE International Symposium on Electromag.

[7]   A. Waizman, C.-Y. Chung, "Resonant Free Power Network Design Using Extended Adaptive Voltage Positioning (EAVP) Methodology", IEEE TRANSACTIONS ON ADVANCED PACKAGING, VOL. 24, NO. 3, pp. 236-244, 2001.

[8]   B. Archambeault, "The Effect of Decoupling Capacitor Distance on Printed Circuit Boards Using Both Frequency and Time Domain Analysis", 2005 IEEE International Symposium on Electromagnetic Compatibility, pp. 650-654, 2005.

[9]   B. Archambeault, J. Kim, S. Connor, J. Fan, "Optimizing Decoupling Capacitor Placement to Reduce Effective Inductance", 2011 IEEE International Symposium on Electromagnetic Compatibility, pp. 179-183, 2011.

[10] F. Carrió, V. González, E. Sanchis, D. Barrientos, J.M. Blasco, F.J. Egea, "A Capacitor Selector Tool for On-Board PDN Designs in Multi Gigabit Applications", 2011 IEEE International Symposium on Electromagnetic Compatibility, pp. 367-372, 2011.

Send Orders for Reprints to reprints@benthamscience.net
Switching Mode Circuit Analysis and Design, 2013, 39-56   **39**

CHAPTER 4

Validation of Decoupling Performance of Capacitor by PCB [1- 3]

Abstract: The measurement result of the performance of the decoupling capacitors which were used on several redesigned MB of the commercialized workstation is presented.

Keywords: Decoupling performance, workstation, mother board (MB), meander trace, radiated electric field (EF) strength, common mode (CM) current, magnetic field (MF) strength, three terminal capacitors, ferrite beads, PDN layer, immunity test, signal trace (ST) layer, ground plane (GP) layer, power plane (PP) layer, power trace (PT) layer, spectra, I/O signal cables, current probe, magnetic near-field, semi anechoic chamber.

EMI REVIEW OF REDESIGNED MB OF FIRST WORKSTATION

When the characteristic impedance of the transmission line connected to the decoupling capacitor is increased, S_{21} will reduce or the decoupling performance will increase. The limit of the decoupling performance by using the capacitor was validated on the commercialized workstation by this way. This way cannot be applied straightly to the actual SMC because the risk which increases the terminal impedance at the power terminal of LSI exists.

Fig. **1a** shows the MB of the first workstation (EWS4800/320: NEC) for the validation of the limit of the performance of the decoupling by the capacitor. Fig. **1b** shows a part of the redesigned pattern of the PDN layer of the MB which was redesigned it of the first workstation.

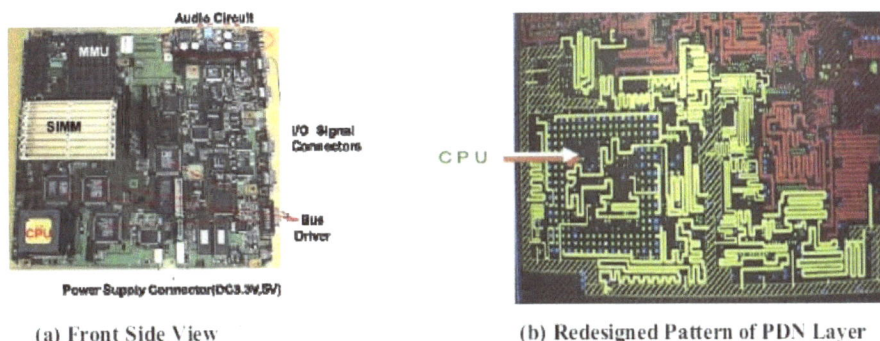

(a) Front Side View   (b) Redesigned Pattern of PDN Layer

Figure 1: MB of First Workstation.

In Fig. **1a**, SIMM means the connector area of the single inline memory module, PCB consisted of 4 layers which were the signal trace (ST) layer, the ground plane (GP) layer, the power plane (PP) layer, the ST layer, and the system clock frequency was 40MHz. In Fig. **1b**, PDN consist of the meander traces of 1mm width and the stem traces, the meander traces are connected to the power terminal of IC/LSI, and the stem traces are connected between the meander traces and the power receiving terminal, PCB consisted of 6 layers which were the ST layer, the GP layer, the power trace (PT) layer, the PT layer, the GP layer, and the ST layer. The microstrip transmission line consisted by the meander trace and the ground plane has the relatively high characteristic impedance. On the other hand, the microstrip transmission line consisted by the stem trace and the ground plane has the relatively low characteristic impedance. The chip-ceramic capacitors are used to the meander traces nearby LSI. The aluminum capacitors having the large capacitance are used to the stem traces. All signal patterns were not changed of the redesigned MB.

The redesigned MB was prototyped and housed in the first workstation. The first workstations housed the original MB and the redesigned MB with the CRT monitor, the keyboard, and the mouse are measured the radiated (EF) strength in the formal test site at the 10m distance.

Fig. **2** shows the measured strength of the radiated EF of the first workstation.

Frequency (MHz)	Original		Redesigned		Effect	
	Vertical	Horizontal	Vertical	Horizontal	Vertical	Horizontal
	(dBμV/m)		(dBμV/m)		(dB)	
80	42	38	37		-5	-12
320	32	33	26		-6	-4
360	37	.	32	.	-5	.
480	41	.	33	.	-8	.
800	45	.	39	.	-6	.
920	40	.	36	.	-4	.

Figure 2: Measured Strength of Radiated EF.

In Fig. **2**, the largest 6 spectra which are the multiple of the 40*MHz* of the system clock frequency are presented, and the strength of the vertical radiated electric field was suppressed from 4dB to 8dB.

Fig. **3** shows the setup of the MB of the first workstation for measuring the common mode (CM) current of I/O signal cables.

Figure 3: Setup for CM Current of I/O Signal Cables.

In Fig. **3**, the prototyped current probe and HP8563E for the spectrum analyzer were used. It has been believed that the measure cause of the radiated emission is CM current on the I/O signal cables. Two RS232C cables, a printer cable, a keyboard cable, and a mouse cable were connected to MB and they were measured by the commercialized current probe. In addition, CM current of the keyboard cable and the mouse cable were measured as a group.

Fig. **4** shows the measured CM current of the I/O signal cables.

Frequency [MHz]	Original				Redesigned				Effect			
	RS23 2C#1	RS23 2C#2	Priter cable	KB/ Mouse	RS23 2C#1	RS23 2C#2	Priter cable	KB/ Mouse	RS23 2C#1	RS23 2C#2	Priter cable	KB/ Mouse
	(dBuV)				(dBuVm)				(dB)			
80	33	33	37	39	30	30	25	25	-3	-3	-12	-14
320	52	47	50	48	48	43	45	42	-6	-4	-5	-6
360	41	42	42	42	41	41	31	31	0	-1	-11	-11
480	54	54	51	54	40	41	44	44	-14	-13	-7	-10
800	43	33	44	41	45	45	39	36	2	12	-5	-5
920	49	46	45	45	45	44	39	38	-4	-2	-6	-7

Figure 4: Measured CM Current of I/O Signal Cables.

In Fig. **4**, the measured values show the peak voltage of between the terminals of the conventional current probe, these values can be only used for comparison in Fig. **4**, and CM current is also suppressed from 0dB to 14dB.

Fig. **5** shows the measurement system of the magnetic near-field. The magnetic probe was held on the head of the scanner which was made of the polyvinyl chloride (PVC) for forming the free space, and the head was scanned to the x-axis and the y-axis.

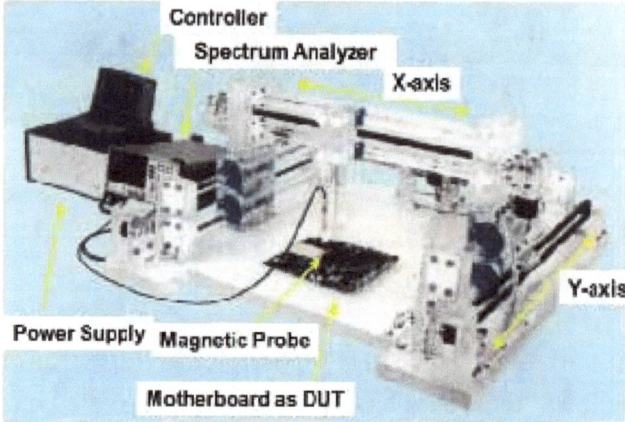

Figure 5: Measurement System of Magnetic Near-Field.

Fig. **6** shows the measured distribution of the MF on the surface of the original MB.

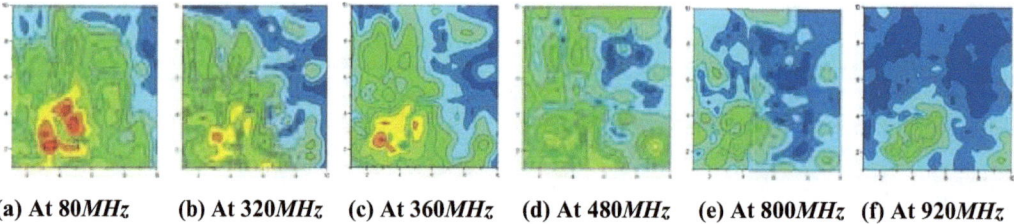

(a) At 80*MHz*   (b) At 320*MHz*   (c) At 360*MHz*   (d) At 480*MHz*   (e) At 800*MHz*   (f) At 920*MHz*

Figure 6: Distribution of MF on Surface of Original MB.

Fig. **7** shows the measured distribution of MF on the surface of the redesigned MB.

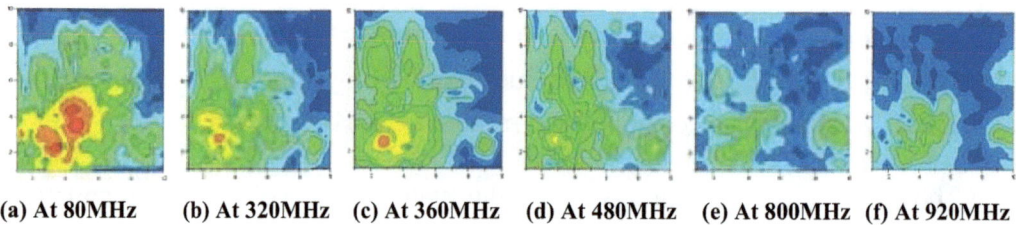

(a) At 80MHz   (b) At 320MHz   (c) At 360MHz   (d) At 480MHz   (e) At 800MHz   (f) At 920MHz

Figure 7: Measured Distribution of MF on Surface of Redesigned MB.

In Figs. **6** and **7**, strongest area of MF is around CPU and the weakest area of MF is around the audio circuit. MF strength up to $480MHz$ was reduced around the area of the I/O signal connectors of the redesigned MB.

Fig. **8** shows the scale marker of MF strength in $dB\mu A/m$.

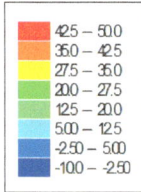

42.5 – 50.0	440 – 500
35.0 – 42.5	380 – 440
27.5 – 35.0	320 – 380
20.0 – 27.5	260 – 320
12.5 – 20.0	200 – 260
5.00 – 12.5	140 – 200
-2.50 – 5.00	80.0 – 140
-10.0 – -2.50	20.0 – 80.0
	-4.00 – 20.0
	-10.0 – -4.00

(a) For Figure 7a   **(b) For Except Figure 7a**

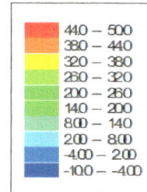

Figure 8: Scale Marker of MF Strength.

EMI REVIEW OF REDESIGNED MB OF SECOND WORKSTATION

Fig. **9** shows the original MB of the second workstation (EWS4800/420: NEC). In Fig. **9**, PCB of the original MB consisted of 10 layers which were the ST layer, the GP layer, the ST layer, the ST layer, the PP layer, the ST layer, and the ST layer. The system clock frequency was $66.66MHz$. The power layer was redesigned by forming the meander traces and the stem traces by the similar way to the MB of the first workstation. As the result, PCB of the first redesigned MB consisted of 12 layers which were the ST layer, the GP layer, the ST layer, the ST layer, the GP layer, the PT layer, the PT layer, the GP layer, the ST layer, the ST layer, the GP layer, and the ST layer.

Figure 9: Original MB of Second Workstation.

Fig. **10** shows the configuration of the second workstation which was placed in the semi anechoic chamber for the measurement of the radiated EF strength and the T&D program was running. In Fig. **10a**, the cables for the mouse, the keyboard, and the CRT display were connected. In Fig. **10b**, only the CRT display was connected. In Fig. **10c**, the cables for the mouse, keyboard, 10base internet, NPCI2 interface, and printer were connected to the workstation but the other equipment were not connected to these cables.

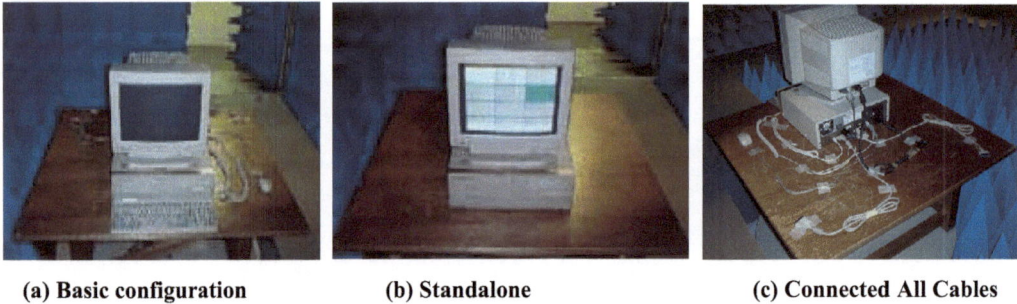

| (a) Basic configuration | (b) Standalone | (c) Connected All Cables |

Figure 10: Configuration of Second Workstation in Semi Anechoic Chamber.

Fig. **11** shows the measured strength of the radiated EF of the horizontal polarization of the original MB and the first redesigned MB housed in the second workstation at the configuration of Fig. **10a**. In Fig. **11**, the measured strength of the radiated EF had become increased by redesigning the MB.

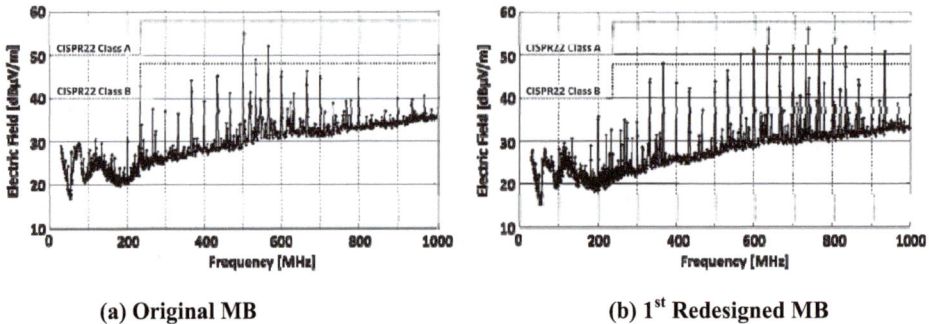

| (a) Original MB | (b) 1st Redesigned MB |

Figure 11: Measured Strength of Radiated EF of 2nd Workstation at Configuration of Fig. **10a**.

Fig. **12** shows the measured strength of the radiated EF of the horizontal polarization of the original MB and the first redesigned MB housed in the second workstation at the configuration of Fig. **10b**. In Fig. **12**, the radiated EF strength in the case of the second workstation housed the original MB is larger than it in the case of the second workstation housed the first redesigned MB up to $700 MHz$ approximately.

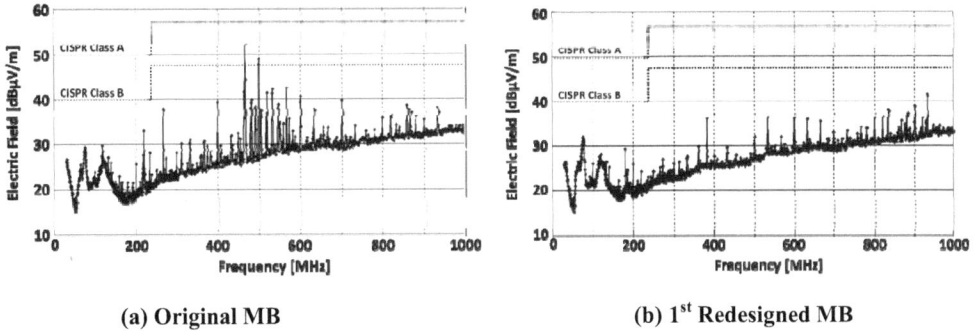

(a) Original MB   (b) 1st Redesigned MB

Figure 12: Measured Strength of Radiated EF of 2nd Workstation at Configuration of Fig. **10b.**

Fig. **13** shows the measured strength of the radiated EF of the horizontal polarization of the original MB and the first redesigned MB housed in the second workstation at the configuration of Fig. **10c.**

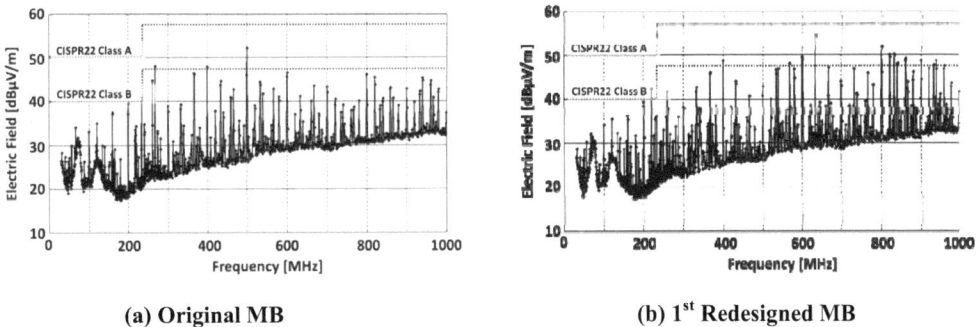

(a) Original MB   (b) 1st Redesigned MB

Figure 13: Measured REF Strength of 2nd Workstation at Configuration of Fig. **10c.**

In Fig. **13**, the radiated EF strength of the second workstation housed original MB is larger up to approximate $700 MHz$ than it when the first redesigned MB was housed.

Fig. **14** shows the MB of the second workstation, which was placed in the semi anechoic chamber for the measurement of the radiated EF strength and the T&D program was running.

(a) Standalone   (b) Connected All Cables

Figure 14: Configuration of MB of Second Workstation in Semi Anechoic Chamber.

Fig. **15** shows the measured strength of the radiated EF of the horizontal polarization of the original MB and the first redesigned MB at the configuration of Fig. **14a**.

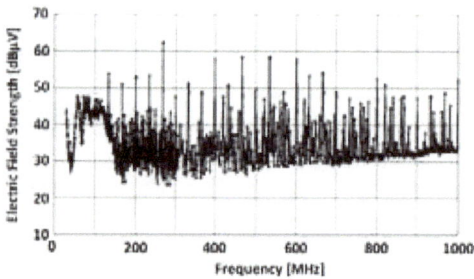

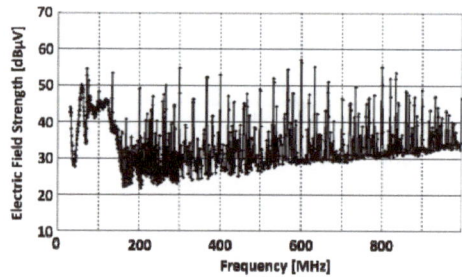

(a) Original MB   (b) 1st Redesigned MB

Figure 15: Measured Strength of Radiated EF of MB at Configuration of Fig. **14a**.

In Fig. **15**, the radiated EF strength of the original MB is larger than it of the first redesigned MB up to $600MHz$. However, it is the opposite over the $600MHz$.

Fig. **16** shows the measured strength of the radiated EF of the horizontal polarization of the original MB and the first redesigned MB at the configuration of Fig. **14b**. In Fig. **16**, the radiated EF strength of the first redesigned MB seems to be larger than it of the original redesigned MB slightly up to $1GHz$.

From above measurement result, it was discussed about the test result which shows that the radiated EF strength of the first redesigned MB is larger than it of the original MB, and the layout of the signal traces of the original MB and the first redesigned MB which were almost same was reviewed. As the result, the following was clarified; the I/O signal traces of the original MB were placed nearby the clock signal traces and CPU and other high-speed LSI, the radiated EF strength of the original MB was relatively large, and the increased MF nearby CPU and other high-speed LSI interfered to the I/O signal trace.

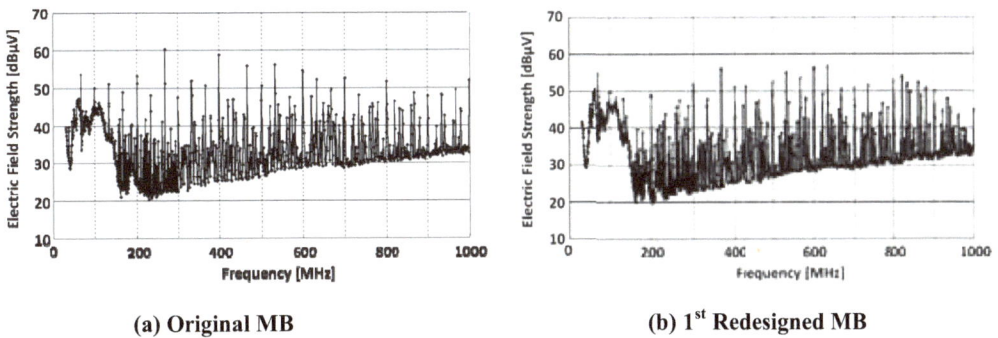

(a) Original MB   (b) 1st Redesigned MB

Figure 16: Measured Strength of Radiated EF of MB at Configuration of Fig. **14b**.

Fig. **17a** shows the front view of the improved MB. Fig. **17b** shows the layout of the devices and the signal traces of the original MB. Fig. **17c** shows the layout of the improved MB.

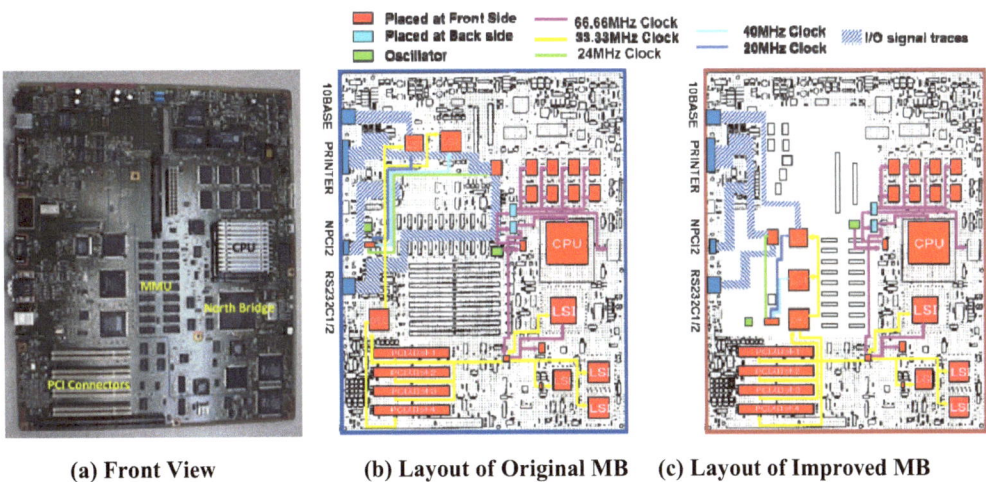

(a) Front View   (b) Layout of Original MB   (c) Layout of Improved MB

Figure 17: Improved MB.

In Fig. **17**, the SIMM connectors which were confirmed to be no influence to EMI were removed and 18 pieces of MMU were shift to the space of the SIMM connectors, the vacant space was used for arrangement of the I/O signal and clock signal traces, the other signal traces were not changed, the layer formation of PCB was same as the first redesigned MB. 6 types of formation of the PT layer were validated by the improved MB. The typical type-A of the PT layer was redesigned by forming the meander traces and the stem traces by the similar way to the MB of the first workstation. The typical type-B of the PT layer was redesigned the type-A by minimizing the stem traces.

Fig. **18** shows measured strength of the radiated EF of the horizontal polarization of the type-A and type-B of the improved MB housed in the second workstation at the configuration of Fig. **10c**.

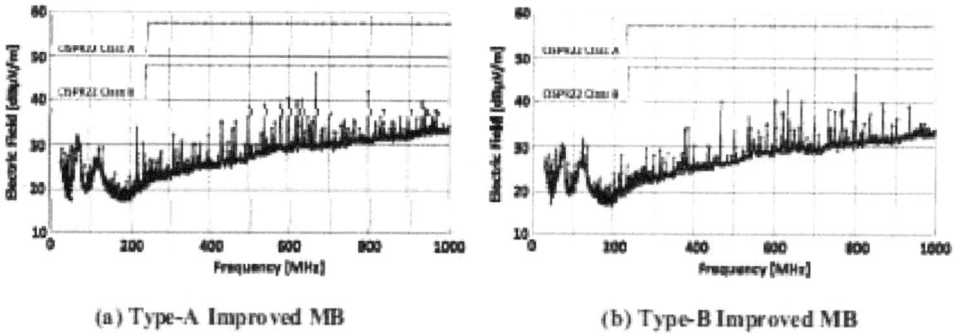

(a) Type-A Improved MB   (b) Type-B Improved MB

Figure 18: Measured Strength of the Radiated EF of 2nd Workstation at Configuration of Fig. **10c**.

In Fig. **18**, the spectra in the case of the type-A improved MB were reduced 10 or 15 dB against it of the original MB.

Fig. **19** shows measured strength of the radiated EF of the horizontal polarization of the type-A and type-B of the improved MB at the configuration of Fig. **10b**.

In Fig. **19**, the spectra in the case of the type-A improved MB were smaller than it of the type-B improved MB, and the type-A improved MB was got the best result.

Fig. **20** shows the measured strength of the radiated EF of the horizontal polarization of type-A improved MB and the type-A improved MB at the configuration of Fig. **14a**.

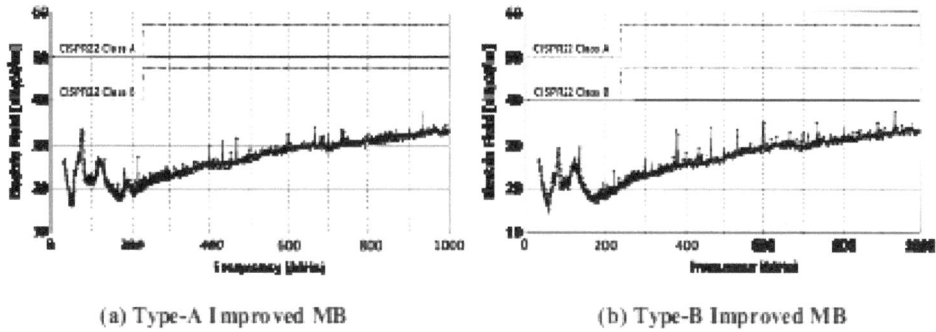

(a) Type-A Improved MB   (b) Type-B Improved MB

Figure 19: Measured Strength of Radiated EF of 2nd Workstation at Configuration of Fig. 10b.

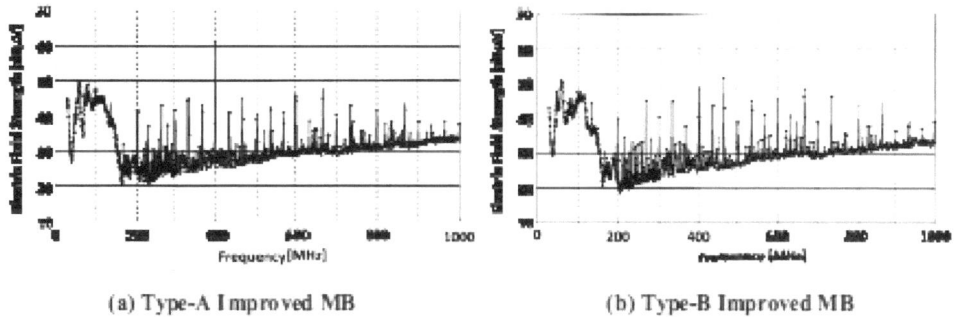

(a) Type-A Improved MB   (b) Type-B Improved MB

Figure 20: Measured Strength of Radiated EF of 2nd Workstation at Configuration of Fig. 14a.

Fig. 21 shows the measured strength of the radiated EF of the horizontal polarization of the original MB and the first redesigned MB at the configuration of Fig. 14b.

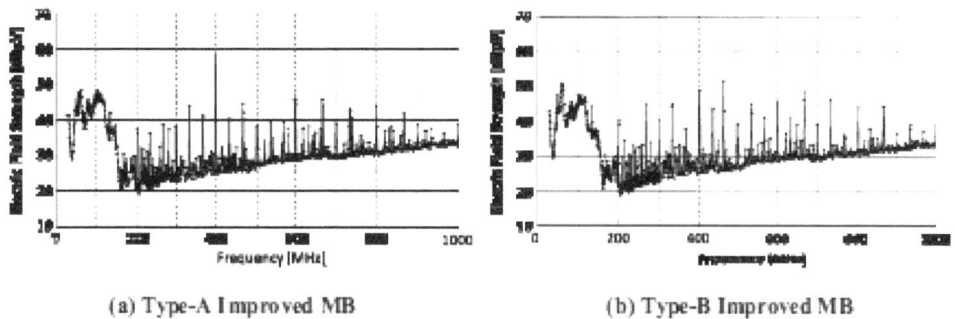

(a) Type-A Improved MB   (b) Type-B Improved MB

Figure 21: Measured Strength of Radiated EF of 2nd Workstation at Configuration of Fig. **14b**.

Fig. **22** shows the measured distribution of MF on the surface of the second workstation at 266.6*MHz*. Fig. **22a** shows it of the original MB, Fig. **22b** shows it of the type-A improved MB, and Fig. **22c** shows it of the type-B improved MB.

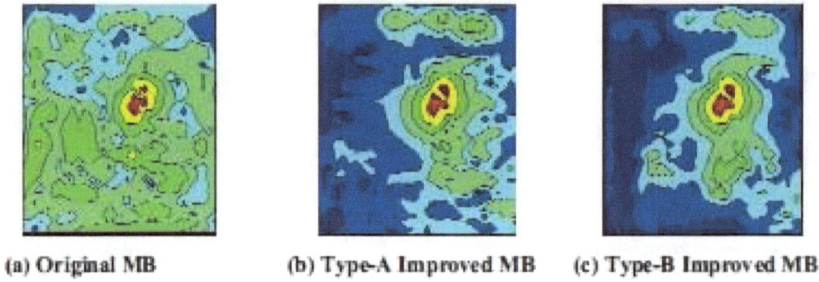

(a) Original MB   (b) Type-A Improved MB   (c) Type-B Improved MB

Figure 22: Measured Distribution of MF on Surface of 2nd Workstation.

In Fig. 22, the strongest magnetic field exists nearby CPU. The signal traces of the I/O interface were placed at area of the right side of CPU, the magnetic field at this area is seemed to be weakest, and this fact is dovetailed with the measurement result about the radiated electric field.

Fig. 23 shows the comparison of the measured strength of the radiated EF between the original MB and the type-A improved MB, and the effectiveness of the EMC suppressing component such as the three terminal capacitors and the ferrite beads for using the I/O signal traces. The radiated electric field strength was increased by removing the EMC suppressing components in the case of the original MB, and the EMC suppressing components has no effect in the case of the type-A improved MB. From above, it was clarified that the signal traces on MB are interfered by the EMW on the PDN greatly.

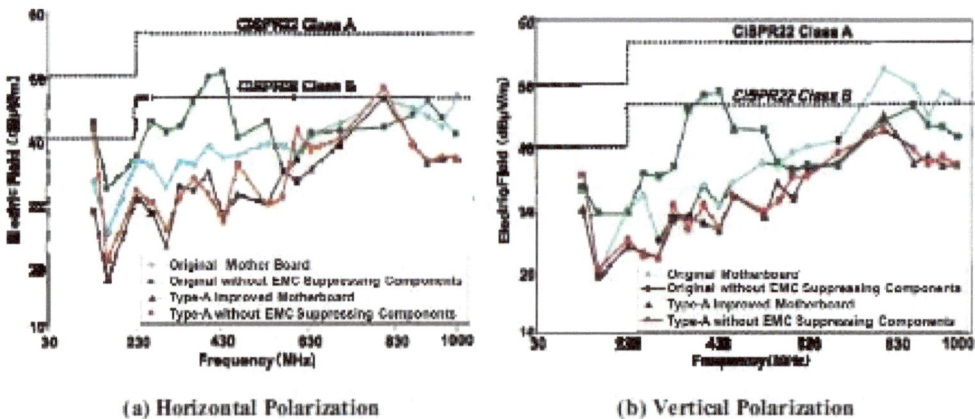

(a) Horizontal Polarization   (b) Vertical Polarization

Figure 23: Measured Strength of Radiated EF of 2nd Workstation.

IMMUNITY REVIEW OF REDESIGNED MB OF SECOND WORKSTATION

The testing and the measurement techniques of Immunity of the equipment under test (EUT) are being standardized by IEC 61000-4. The concerning standards of the sub number of IEC61000-4 of the enhanced decoupling circuit are -2 presents about the electrostatic discharge immunity test, -3 presents about the radiated, the radio-frequency, and the electromagnetic field immunity test, -4 presents about the electrical fast transient/burst immunity test, -5 presents about the surge immunity test, -6 presents about Immunity to conducted disturbances, induced by radio-frequency fields.

The second workstation housed the original and improved type-A.

Fig. **24** shows the test condition.

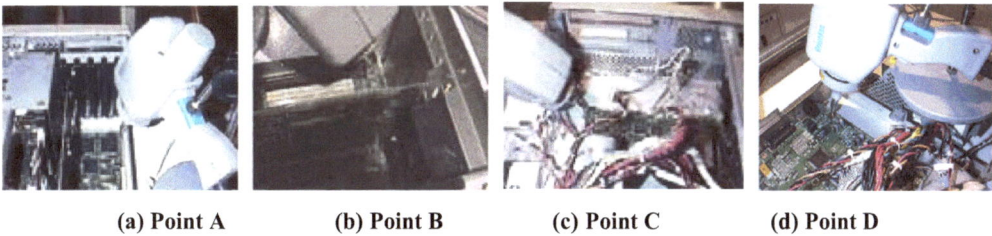

| (a) Point A | (b) Point B | (c) Point C | (d) Point D |

Figure 24: Test Condition of ESD Immunity Test.

	Test Voltage (kV)	EUT	
		Original MB	Type-A improved MB
Point A	±4.0	Pass	Pass
	+4.5	Pass	Error of SCSI #0
	−5.0	Error of SCSI#0, KeyBoard	-
Point B	±4.0	Pass	Pass
	+4.5	Error of SCSI #0	Pass
	−4.5	-	Pass
	±7.0	-	Pass
	+7.5	-	Pass
	−7.5	-	Error of SCSI#0
Point C	±3.5	Pass	Pass
	−4.0	Error of SCSI #0	Pass
	+4.0	-	SHUT DOWN
Point D	±3.0	Pass	Pass
	+3.5	Error of SCSI #0	Pass
	−3.5	-	Pass
	+4.5	-	Pass
	−4.5	-	Pass
	+5.0	-	Error of SCSI #0, SCSI

Figure 25: Test Result of ESD Immunity Test.

In Fig. **24**, the ESD test was done by touching the head of the ESD gun to the attachment screw of MB in the workstation because the formal test result was passed. The ESD immunity was improved greatly at the point B and D. The I/O interface signal traces existed at the area between the point B and D.

Fig. **25** shows the test result of the ESD immunity test.

Fig. **26** shows the test condition.

Figure 26: Test Condition of Radiated, Radio-frequency, and Electromagnetic Field Immunity Test.

In Fig. **26**, the test was done by removing the enclosure of the second workstation.

Fig. **27** shows the test result of the radiated, the radio-frequency, and the electromagnetic field immunity test.

Test EF Strength (V/m)	EUT	
	Original MB	Type-A Improved MB
30	Pass	Pass
35	Error of SCSI #0 at 799.9MHz	Pass
36.8	Error of SCSI #0 at 966.6MHz	Pass
37	Ditto	Shutdown at 200MHz Error of SCSI #0 at 799.9MHz
40	Shutdown at 466.6MHz	

Figure 27: Test Result of Radiated, Radio-frequency, and Electromagnetic Field Immunity Ttest.

Fig. **28** shows the test condition.

(a) Power Line Test

(b) Signal Line Test

Figure 28: Test Condition of Electrical Fast Transient/Burst Immunity Test.

In Fig. **28**, the electrical fast transient/burst test was done by the formal method.

Fig. **29** shows the test result of the electrical fast transient/burst immunity test.

Target	Test Voltage (kV)		EUT	
			Original MB	Type-A Improved MB
Power Line	-3.0		Error of Keyboard	Error of Keyboard
Signal Line	RS232C1	+1.8	Error of Keyboard	Pass
		+1.5	-	Error of Keyboard
	10BASE	+3.6	Pass	Pass
		+3.7	Error of Keyboard	Pass
		+4.5	-	Pass
		-3.3	Pass	Pass
		-3.4	Error of Keyboard	Pass
		-4.1	-	Pass
		-4.2	-	Error of Ethernet

Figure 29: Test Result of Electrical Fast Transient/Burst Immunity Test.

Fig. **30** shows the test condition of the surge immunity test.

Figure 30: Test Condition of Surge Immunity Test.

In Fig. **30**, the surge Immunity test was done by the formal method.

Fig. **31** shows the test result of the surge immunity test.

Voltage (kV)		EUT	
		Original MB	Type-A Improved MB
Between Lines	+2.0	Pass	Pass
Between Line and Ground	+2.5	Pass	Pass
	+2.7	Pass	Pass
	+3.0	Pass	Pass

Figure 31: Test Result of Surge Immunity Test.

Fig. **32** shows the test condition of Immunity to conducted disturbances, induced by radio-frequency fields.

Figure 32: Test Condition of Immunity to Conducted Disturbances, Induced by Radio-Frequency Fields.

In Fig. **32**, the test of Immunity to conducted disturbances, induced by radio-frequency fields was done by the formal method.

Fig. **33** shows the test result of Immunity to conducted disturbances, induced by radio-frequency fields.

Voltage (kV)		EUT	
		Original MB	Type-A Improved MB
10	RS232C1	Pass	Pass
	10BASE	Pass	Pass

Figure 33: Test Result of Immunity to Conducted Disturbances, Induced by Radio-Frequency Fields.

In Fig. **33**, the value of 10kV was the maximum voltage of the instrument.

Fig. **34** shows the test result of the high voltage pulse immunity test.

Voltage (kV)	Pulse Width (ns)	Target Line	EUT	
			Original MB	Type-A Improved MB
±0.6	800	Hot, Neutral	Pass	
	50	Ditto	Pass	Pass
+0.7	800	Ditto	Pass	Pass
	50	Ditto	Pass	Pass
-0.7	800	Hot	Pass	Pass
		Neutral	Error of Keyboard	Pass
	50	Hot, Neutral	-	Pass
±1.2	800	Ditto	-	Pass
	50	Ditto	-	Pass
±1.3	800	Ditto	-	Pass
	50	Ditto	-	Pass
-1.3	800	Ditto	-	Pass
	50	Neutral	-	Pass
		Hot		Error of Keyboard

Figure 34: Test Result of High Voltage Puse Immunity Test.

In Fig. **34**, immunity level of the second workstation housed the type-A improved MB was improved greatly, the high voltage pulse immunity test was developed by IBM, and this test method had been used to the mainframe of the computer. However this method is not standardized by IEC.

CONCLUSIONS

From above validation, the following facts were clarified; the almost leakage of the EMW or EMI from ITE depend on the decoupling circuit of PDN of PCB, effective limit of suppressing EMI by the decoupling performance of the capacitor was considered to be 15dB, and the enhancement of the decoupling circuit is effective for improving immunity of ITE.

REFERENCES

[1]   S. Yoshida, H. Tohya, "Novel decoupling circuit enabling notable electromagnetic noise suppression and high-density packaging in a digital printed circuit board" IEEE International Symposium on Electromagnetic Compatibility Record, pp. 641-646, 1998.

[2]   H. Tohya, "New Technologies Doing Much For Solving the EMC Problem in the High Performance Digital PCBs and Equipment" IEICE Transactions on Fundamentals of Electronics, Communications and Computer Sciences, Vol. E82- A, No. 3, 1999.

[3]   S. Yoshida, H. Tohya, "Novel Decoupling Circuit Comprising Magnetic Materials and Build-in Choking Coils" IEICE / IEEJ / IEEE International Symposium on Electromagnetic Compatibility (EMC '99 TOKYO) Record, pp. 616-619, 1999.

Send Orders for Reprints to reprints@benthamscience.net
Switching Mode Circuit Analysis and Design, 2013, 57-85 57

CHAPTER 5

Conformation of EMI Level and Suppressing Method of LSI

Abstract: EMC of PCB and ITE depends on the decoupling performance of PDN. The clarification of the relationship between the power current of LSI and the EMI of PCB became necessary.

Keywords: EMI, FR-4 board, magnetic probe, SMA connector, calibration factor, network analyzer, MP method, power current, IEC61967-6, LSI, ASIC, radiated EF strength, signal current, I/O interface, core circuit block, I/O circuit block, clock frequency, on-chip capacitor, microprocessor, capacitor cell.

MAGNETIC PROBE

The major active component on PCB will be LSI. LSI consists of large amount of transistor. The transistor generates EMW for forming the signal. The effective decoupling circuit of the power supply wiring has not been formed in LSI. On the other hand, the signal line is isolated electromagnetically by the on-chip inverter in LSI. Therefore, it was considered that the leaked EMW into the power line of PCB will be the major cause of EMI on PCB and the development of the magnetic probe for measuring the high-frequency power current was started.

The magnetic probe was designed in accordance with the following conditions;

a. To be able to probe the power current of LSI easily and accurately on PCB.

b. To be non-contact measurement.

c. To be able to probe at the wide frequency band.

Fig. **1** shows PCB used in the magnetic probe. Each Fig. **1a** and **1b** shows PCB and the formation of PCB. The diameter of PCB was $10mm \times 50mm$, PCB of the magnetic probe was manufactures by the 5 metal-layer FR-4 board, A shows the area for SMA connector, B shows the strip conductor of the 50Ω microstrip line, and C shows the sensor part and the diameter of the square window was $8.4mm \times 0.2mm$. In Fig. **1b**, the sensor part consists of the strip conductor layer and two copper plane layers, the magnetic flux passes through the square window of the copper plane, the diameter of the square window was *2mm×0.2mm*, and the vias were used for the cancellation of the electric field.

(a) PCB

(b) Formation of Sensor Part

Figure 1: First Developed Magnetic Probe.

Fig. **2** shows the commercialized magnetic probe which consists of PCB shown in Fig. **1a**. Each Fig. **2a** and **2b** shows the product and the setup for probing the spectra of the power current. The magnetic probe was commercialized in end of 1998, PCB was covered by the plastics, and SMA connector was equipped. In Fig. **2b**, the magnetic probe was set on the power distribution line on the test board by using the fixture [1].

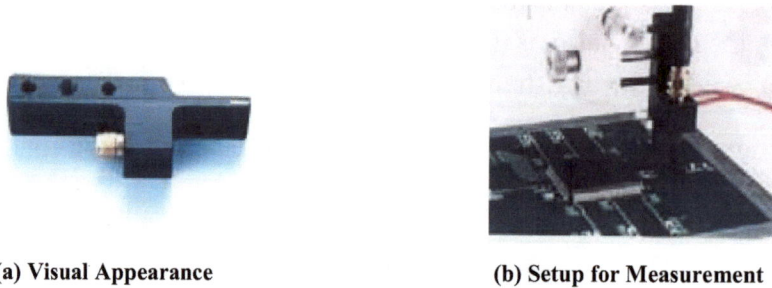

(a) Visual Appearance

(b) Setup for Measurement

Figure 2: Commercialized Magnetic Probe.

The electric current is got by

$$I_{dB} = V_{p\ dB} + C_{f\ dB} - C_{h\ dB} \tag{1}$$

where V_P is the output voltage of the magnetic probe, and C_f [S/m] is the calibration factor for the magnetic probe, C_h is

$$C_h = \frac{h}{\pi Y(Y+2h)} \tag{2}$$

where h [m] is the thickness of the insulator of the microstrip line, Y [m] is the distance between the strip conductor and the detecting point, and the thickness of the strip conductor is negligibly small.

Fig. **3** shows an example of the calibration factor for the magnetic probe (C_f).

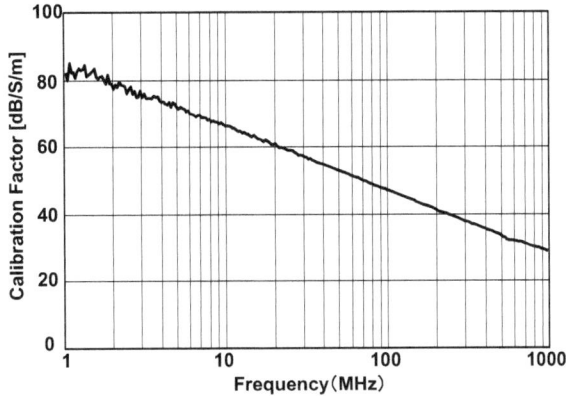

Figure 3: Calibration Factor for Magnetic Probe (C_f).

The square window of PCB having the diameter of $10mm \times 0.2mm$ was also prototyped as the high-sensitivity magnetic probe without the envelope.

Fig. **4** shows the developed magnetic probe measurement system. The proved signal by the magnetic probe is transmitted to the spectrum analyzer (left), the portable PC (right) receives the data from the network analyzer, the spectra of the electric current which is transformed from the magnetic field and processed by the portable PC is being displayed by CRT Monitor.

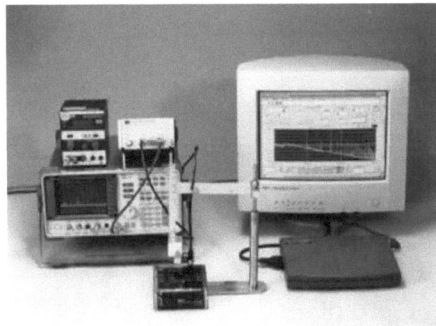

Figure 4: Magnetic Probe Measurement System.

The magnetic probe was supported by the semiconductor engineers because the magnetic probe can test EMI accurately. As the result, measurement method by the magnetic probe was standardized in 2002 as IEC61967-6 which is shown as MP method in "Integrated circuits- Measurement of electromagnetic emissions, 150*kHz* to 1*GHz*" [2]. Fig. **5** shows the improved magnetic probe which was commercialized in September 2000 and standardized in 2008 as Amendment 1 of IEC61967-6. The diameter of the square window for detecting was 1*mm*×0.2*mm* and the insulator of the inner PWB consists of ceramics instead of FR4.

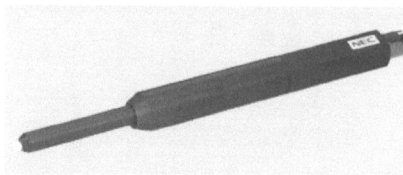

Figure 5: Improved Magnetic Probe.

The evaluation of the spectra of the power current of LSI by MP method was started in 1997 as the project. The quality control engineer in the semiconductor manufacturing sector had been felt confused by EMI claim for a long time from the user such as IT equipment and vehicle equipment manufacturing companies. Traditionally, EMI of LSI was being reduced by the try and error method because the user does not disclose the detail of the test result and the design drawing of PCB and the equipment. Many prototypes of LSI were necessary when this method is used because it is inefficient. The stored prototypes of LSI were selected for DUT of our project.

VALIDATION OF INFLUENCE OF POWER CURRENT OF ASIC TO EMI

ASIC which was manufactured by the 0.55μm rule was selected as the first DUT, because many kinds of ASIC were being used by the vehicle manufacturers and many complaint about EMI were drawn from them.

Fig. **6** shows the setup on LSI tester for the measurement of the spectra of the power current of ASIC. The prototyped high-sensitivity magnetic probe was used for detecting of the spectra of the power current. The voltage of the power supply was $5V$ and $5.5V$. The spectra of the power current of major $5.5V$ for each core and I/O interface circuit block were detected. The spectrum analyzer HP8562E and 25dB amplifier HP8447D were used.

Figure 6: Setup for Measurement of Power Current of ASIC.

Fig. **7** shows the spectra of the ambient noise on the power supply line.

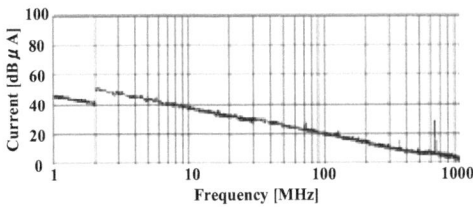

(a) Standby State of LSI Tester   **(b) Operating State of LSI Tester**

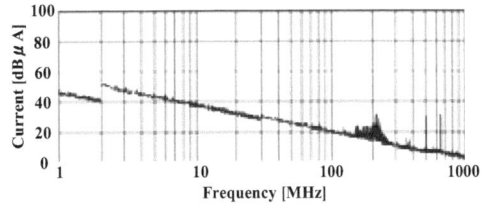

Figure 7: Ambient Noise Spectra on Power Supply Line.

Each Fig. **7a** and **7b** shows the ambient noise at the standby state and the operating state of LSI tester. The original ASIC consisted of the conventional circuit-block. The core circuit-block and the I/O interface circuit-block were not being separated.

Fig. **8** shows measured spectra of the power current of the original ASIC. Each Fig. **8a** and Fig. **8b** shows the spectra of the power current of the core circuit block and the I/O circuit block. The magnitude of the spectra of the power current of the core circuit block and the interface circuit block is similar each and it is relatively large.

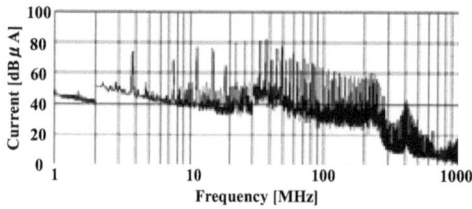

(a) Spectra of Core Circuit Block (b) Spectra of I/O Interface Circuit Block

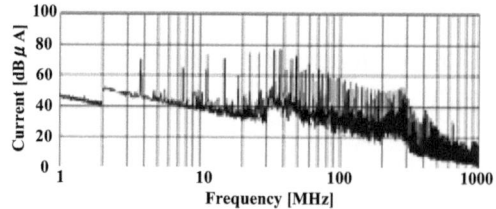

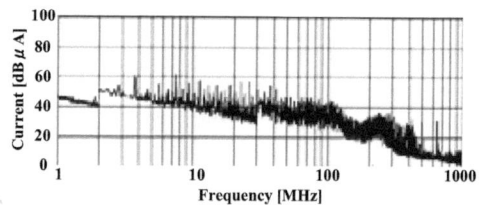

Figure 8: Measured Spectra of Power Current of Original ASIC.

Fig. **9** shows the measured the spectra of the power current of the 1st prototype of ASIC which was consisted of the separated core circuit block and I/O interface circuit block. Each Fig. **9a** and Fig. **9b** shows the spectra of the power current of the power supply pin of the core circuit block and the I/O circuit block. The magnitude of the spectra of the power current of the core circuit block is relatively large but the amplitude of the spectra of the power current of I/O interface circuit block is being reduced largely. The 1st prototype got the satisfying EMI test result from the user of this ASIC. Therefore, the emission test of the ASIC mounted on the test board was tried.

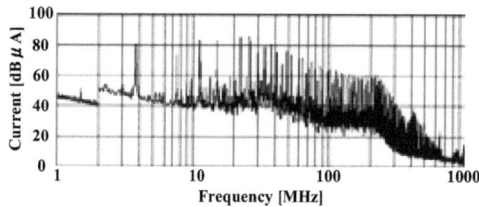

(a) Spectra of Core Circuit Block (b) Spectra of I/O Interface Circuit Block

Figure 9: Measured Spectra of Power Current of 1st Prototype of ASIC.

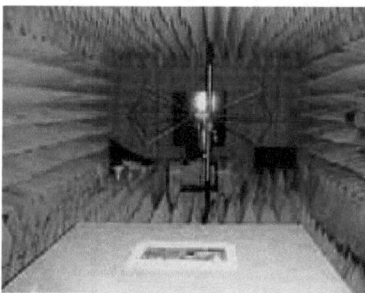

(a) Test Board in Semi-anechoic Chamber (b) Test Board

Figure 10: Setup for Electromagnetic Emission Test of ASIC.

Fig. **10** shows the setup for the electromagnetic emission test of ASIC. Fig. **10a** shows the test board set in the semi-anechoic chamber. Fig. **10b** shows the test board, and the distance between the test board and the antenna was 3*m*. The test board was powered by the dry battery cells on the table.

Fig. **11** shows the measured spectra of the ambient electric field of the semi-anechoic chamber.

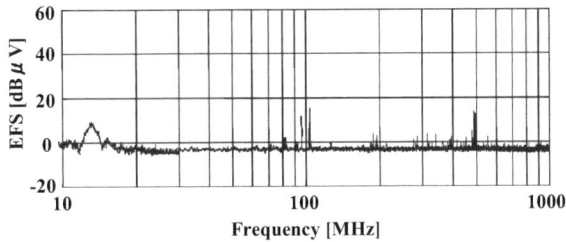

Figure 11: Measured Ambient Electric Field of Semi-anechoic Chamber.

Fig. **12** shows the measured strength of the radiated EF of the test board mounted the original and the first prototype of ASIC.

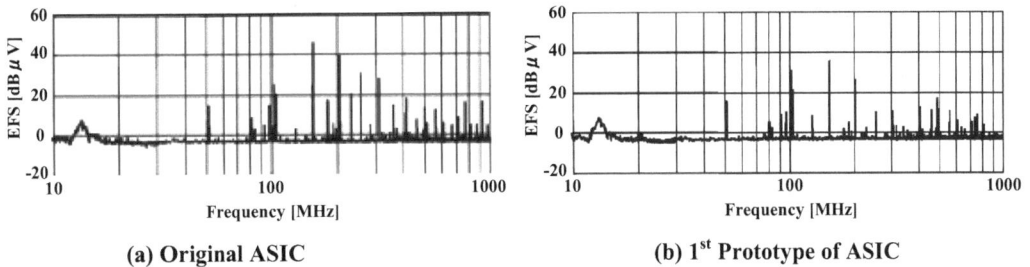

| (a) Original ASIC | (b) 1ˢᵗ Prototype of ASIC |

Figure 12: Measured Strength of Radiated EF of Test Board.

Each Fig. **12a** and **12b** shows the spectra of the test board mounted the original and the first prototype of ASIC. The magnitude of the spectra of the electric field from the test board of the first prototype is being reduced to more than 10dB at over 100*MHz*. This result was considered to depend on the reduced the spectra of the power current on the power distribution line of I/O interface circuit block of the 1ˢᵗ prototype because the core circuit block has not the interface port between PCB except the power distribution line. From above, it was clarified that the spectra of the power current of the PDN on PCB affect EMI greatly.

EMI has been considered to be reduced by loosening the switching period or the rise/fall time in accordance with the conventional idea. In addition, it is well known that the measured spectra of the electric field consist of the spectra of the clock signal almost. The switching period of the second prototype was loosened by increasing the receivers to approximately 400 per clock-tree synthesis driver, which consists of the clock distribution circuit block. The fun-out of the clock-tree synthesis driver of the original and the 1st prototypes was approximately 90. As the result, the clock distribution circuit block was formed by three stages and 5,827 flip-flop circuits were driven by 105 back-stage driver.

Fig. **13** shows the measured spectra of the power current of the second prototype of ASIC. The magnitude of the spectra of the power current of the second prototype was larger than it of each core block of the original ASIC.

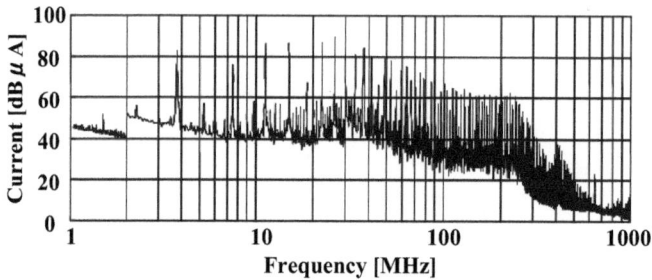

Figure 13: Measured Spectra of Power Current of 2nd Prototype of ASIC.

Fig. **14** shows the measured strength of the radiated EF of the test board of the second prototype of the ASIC. The magnitude of the spectra of the radiated electric field of the test board of the second prototype was larger than it of the 1st prototype. This judgment was matched to it of the user.

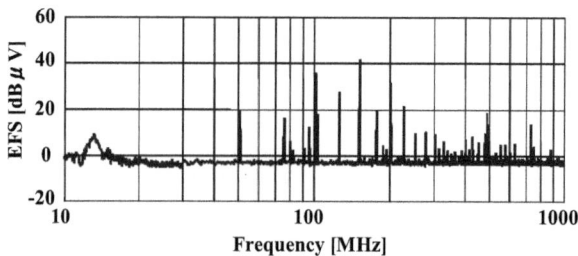

Figure 14: Measured Strength of Radiated EF of Test Board of 2nd Prototype of ASIC.

Fig. **15** shows the simulated signal voltages of the clock driver of the original and second prototype of ASIC. Each Fig. **15a** and **15b** shows the voltage shape of the period of changing ON of the clock driver of the original and second prototype of ASIC. The rise time and the fall time were being loosened to about threefold against it of the original of ASIC.

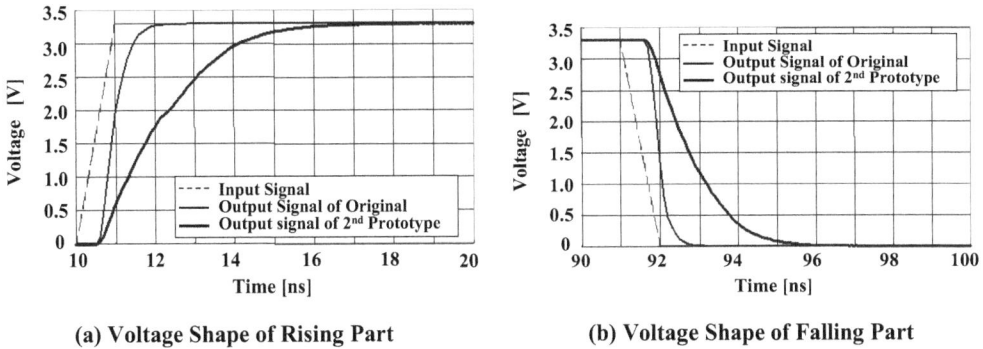

(a) **Voltage Shape of Rising Part**   (b) **Voltage Shape of Falling Part**

Figure 15: Simulated Signal Voltages of Clock Driver of Original and 2nd Prototype of ASIC.

Fig. **16** shows the simulated power current of the clock driver of the original and the second prototype. Each Fig. **16a** and **16b** shows the current shape of the period of changing ON/OFF. The magnitude of the power current is being reduced about 20% against the original however the flowing period of the power current is being increased about threefold. In Fig. **15** and Fig. **16**, SPICE was used, and the parameters were got from the design stage of ASIC. It is also same as the following studies.

Fig. **17** shows the simulated spectra of the power current of second prototype by the Fourier Transform of the power current shown in Fig. **16**.

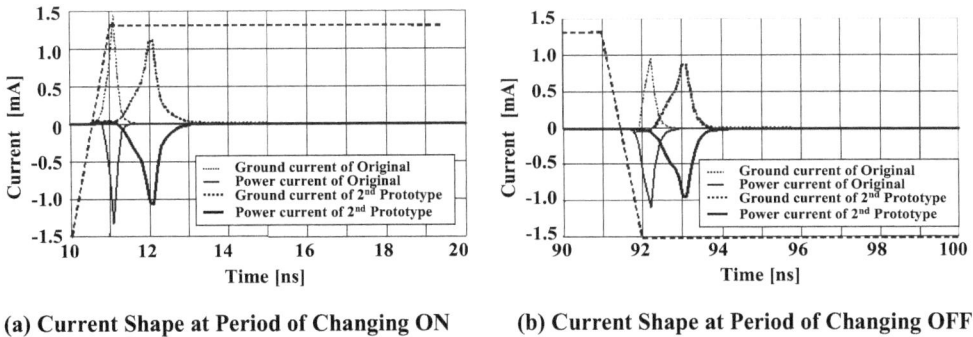

(a) **Current Shape at Period of Changing ON**   (b) **Current Shape at Period of Changing OFF**

Figure 16: Simulated Power Current of Clock Driver of Original and 2nd Prototype of ASIC.

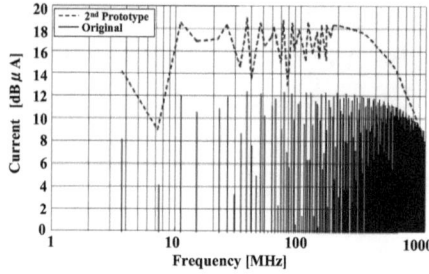

Figure 17: Simulated Spectra of Power Current.

In Fig. **17**, the magnitude of the spectra is smaller than the measured value. The reason is that the measured data shown in Fig. **8** is it of the summation of the current. The tendency of the increase of the magnitude is same between the simulation and the measurement. From above, it was clarified that the EMI suppression effect should not judge by only the slope of the signal voltage.

It had been believed that EMI increases according to the improvement of the technology node of MOSFET, because the radiation efficiency increases by the improvement of the switching speed. In contrast, the power consumption is reduced by the improvement of the technology node. Fig. **18** shows the primitive circuit of clock-tree synthesis of ASIC. Each Fig. **18a** and Fig. **18b** shows the primitive circuits of the original and the third prototype of ASIC. The numeric means the gate length of *nm* unit.

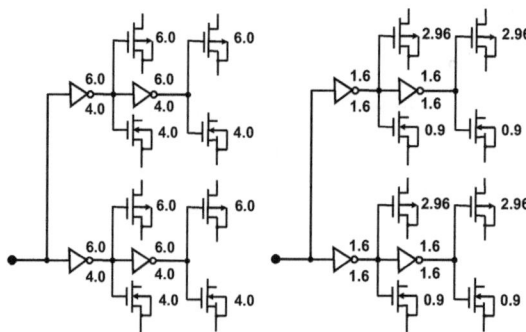

(a) Original ASIC   **(b) 3rd Prototype of ASIC**

Figure 18: Primitive Circuits of clock-tree synthesis.

Fig. **19** shows the measured spectra of the power current of the third prototype of ASIC. The magnitude of the spectra of the power current of the third prototype of

ASIC is slightly smaller than it of the core circuit block of the original, 1st prototype, and second prototype.

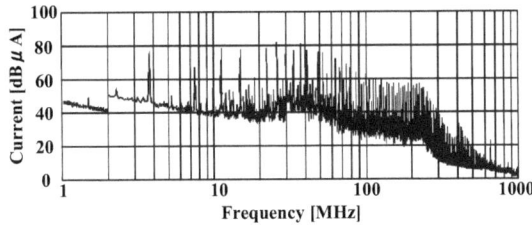

Figure 19: Measured Power Current of 3rd Prototype of ASIC.

Fig. **20** shows the simulated signal voltage at the period of changing ON of the clock driver of the second and third prototype of ASIC.

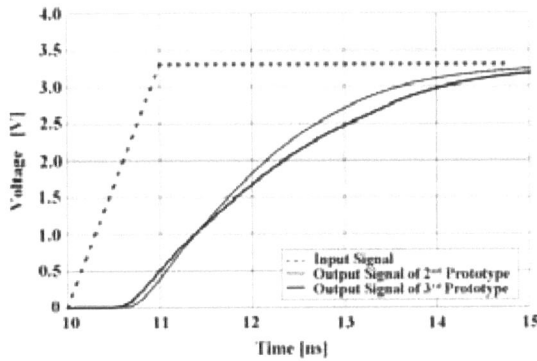

Figure 20: Simulated Signal Voltages at Period of Changing ON of Clock Driver.

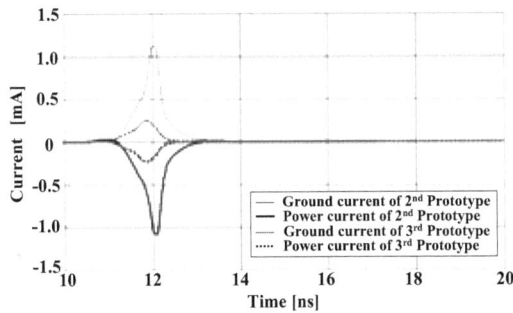

Figure 21: Simulated Power Current at Period of Changes ON of Clock Driver.

Fig. **21** shows the simulated power current at the period of changing ON of the second and third prototype of the clock driver. This waveform is considered to be no relation to the signal voltage and this power current has been hated by the reason that it increases the power consumption and EMI.

Fig. **22** shows the simulated spectra of the third prototype by the Fourier Transform of the power current shown in Fig. **21**.

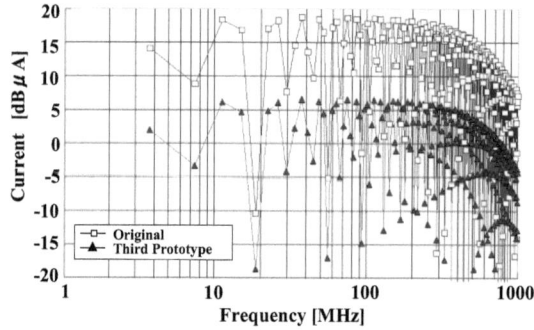

Figure 22: Calculated Spectra of Power Current.

In Fig. **22**, the magnitude of the spectra is smaller than the measured value. The reason is that the measured data shown in Fig. **19** is it of the summation of the current. The difference between the original and the third prototype of the magnitude of the simulated spectra of the power current is 15dB, however the difference between the original and the third prototype of the magnitude of the measured spectra of the power current is 5dB.

Fig. **23** shows the measured spectra of the electric field radiation of the test board of the third prototype of ASIC.

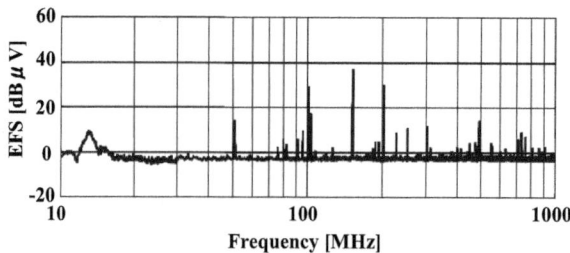

Figure 23: Measured Electric Field Radiation of Test Board of 3rd Prototype of ASIC.

In Fig. **23**, the magnitude of the spectra of the electric field radiation of the test board of the third prototype is 5dB smaller than it of the original and the second prototype of ASIC and it of the third prototype is similar to it of the first prototype. From above, it was clarified that the improvement of the technology node which is defined as the gate length is effective for the EMI suppression.

The technology of the spread spectra clock (SSC) has been applied to MPU and SMPS. It is effective for suppressing EMI because the power of the harmonic waves spread by modulating the clock frequency. The timing of the signals on the clock distribution circuit block of the 4th prototype was divided in order to apply the SSC technology and the clock distribution circuit block was reworked to four groups. The clock-tree synthesis drivers were increased and the fun-out of one clock-tree synthesis driver was reduced to 150 from about 400. The first delay circuit was inserted into the second group, the second delay circuit was inserted into the third group, and the third delay circuit was inserted into the fourth group.

Fig. **24** shows the measured spectra of the power current of the fourth prototype of ASIC. The magnitude of the spectra of the power current of the fourth prototype of ASIC is similar to it of the fifth prototype of ASIC, and it is smaller than it of the others.

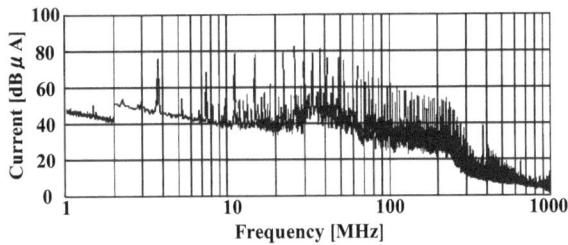

Figure 24: Measured Spectra of Power Current of 4th Prototype of ASIC.

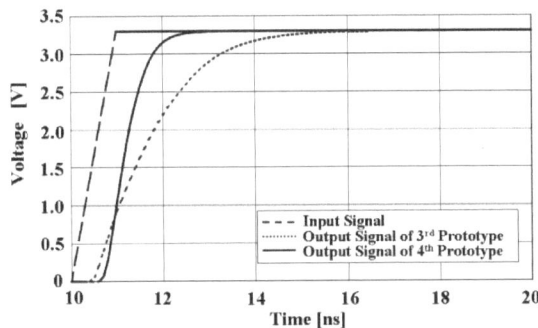

Figure 25: Simulated Power Current at Period of Changing ON of Clock Driver.

Fig. **25** shows the simulated signal voltage at the period of Changes ON of the clock driver of the third and fourth prototype of ASIC.

Fig. **26** shows the simulated power current at the period of changing ON of the clock driver of the third and it of the clock driver of the one group fourth prototype.

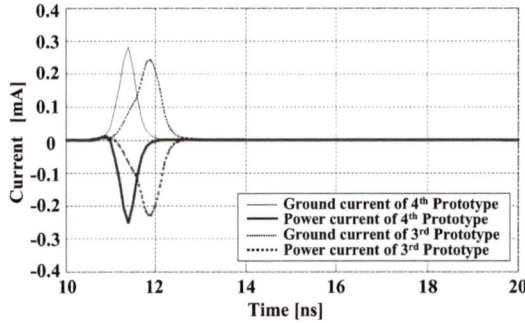

Figure 26: Simulated Power Current at Period of Changing ON of Clock Driver.

Fig. **27** shows the simulated power current shapes at the period of changing ON of the clock driver of fourth prototype. Each #01, #02, #03, #04, and SUM correspond to the power current shape of the first group, second group, third group, forth group, and the summation of the clock driver, respectively.

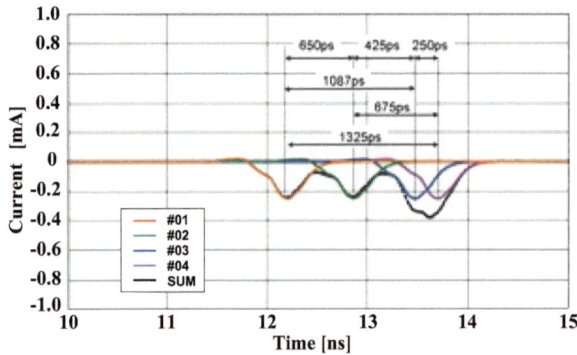

Figure 27: Simulated Power Current at Period of Changing ON of Clock Driver.

Fig. **28** shows the measured spectra of the electric field radiation of the test board of the fourth prototype of ASIC. The spectra of the electric field radiation of the test board of the fourth prototype were smallest in all of prototype of ASIC. The

SSC technology is effective for suppressing EMI because the measured spectra of the electric field were got by the quasi-peak detector in accordance with the current EMC standard of IEC (CISPR16), however performance of ASIC will be reduced. In contrast, the magnitude of the spectra of the electric field is not reduced when it was measured by the peak-hold detector. Now the suppression effect of EMI of the SSC is discussed because the quasi-peak detector depends on the receiver circuit of the AM radio and the digitalization of broadcasting is going now. Therefore, SSC should be applied with based on the awareness of these risks.

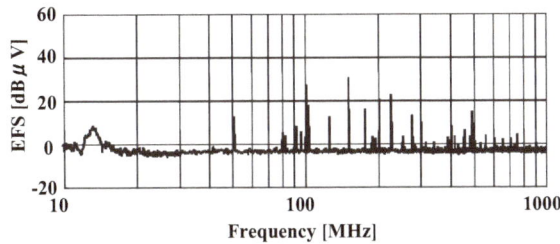

Figure 28: Measured Electric Field Radiation of Test Board of 4th Prototype of ASIC.

POWER CURRENT OF OTHER LSI

Mask ROM

Fig. **29** shows the setup on LSI tester for the measurement of the spectra of the power current of the mask ROM as DUT. The instruments are same as them of ASIC. The voltage of the power supply was $4.6V$ which is the rated voltage of DUT. The access time of DUT was $500ns$. The storage capacity was 8M byte.

Figure 29: Setup for Measurement of Spectra of Power Current of Mask ROM.

Fig. **30** shows the spectra of the ambient noise on the power supply line. Fig. **30a** shows the ambient noise at the standby of LSI tester, Each Fig. **30b** and **30b** shows the ambient noise at the standby state and the operating state of LSI tester.

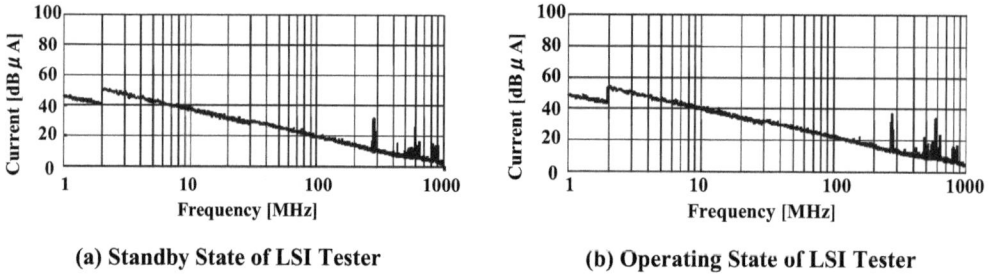

(a) Standby State of LSI Tester   (b) Operating State of LSI Tester

Figure 30: Measured Ambient Spectra of Power Current.

The original mask ROM consisted of the separated core circuit-block and I/O interface circuit-block. The first prototype of the mask ROM was reworked in order to loosen the switching time by reducing the driving performance of I/O buffers. For this objective, each gate width of N-MOSFET and P-MOSFET of the pre-driver was reduced to $5\mu m$ from $15\mu m$ and to $15\mu m$ from $20\mu m$.

Fig. **31** shows the spectra of the power current of I/O interface circuit block of the original and the first prototype of the mask ROM. The magnitude of the spectra of the power current of the first prototype of the mask ROM is smaller at around $100MHz$ than the original of the mask ROM.

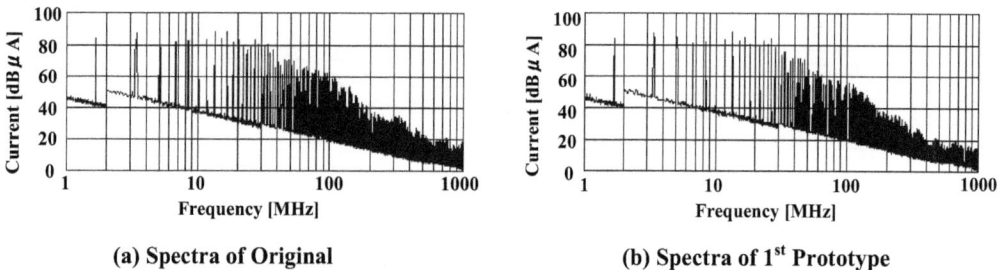

(a) Spectra of Original   (b) Spectra of 1st Prototype

Figure 31: Measured Spectra of Power Current of I/O Interface Circuit Block of Mask ROM.

Fig. **32** shows the measured summation spectra of the power current of I/O interface and core circuit block of the original and the first prototype of the mask ROM. The magnitude of the spectra of the power current of the original and the first prototype of the mask ROM is similar. The first prototype could not get the satisfying EMI test result from the user.

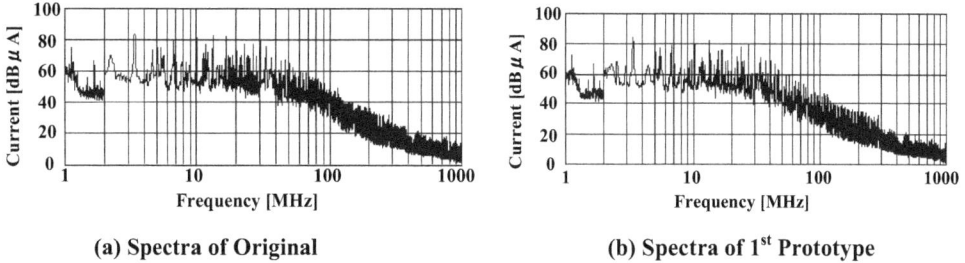

(a) Spectra of Original   (b) Spectra of 1st Prototype

Figure 32: Measured summation Spectra of Power Current of Mask ROM.

The second prototype of the mask ROM was reworked in order to extend the switching period by reducing the driving performance of I/O buffers. For this objective, the resistance of the series resistor of the driver of P-MOSFET was changed to 46.1Ω from 15.4Ω, and resistance of the series resistor to N-MOSFET of the driver was changed to 10.4Ω from 7.8Ω.

Fig. **33** shows the measured spectra of the power current of I/O interface circuit block and the summation spectra of the power current of I/O interface and core circuit block of the second prototype of the mask ROM. The magnitude of the spectra of the power current of I/O interface circuit block and the summation spectra of the power current of I/O interface and core circuit block of the second prototype of the mask ROM are 5-10dB smaller than these of the first prototype of the mask ROM. The second prototype could get the satisfying EMI test result from the user.

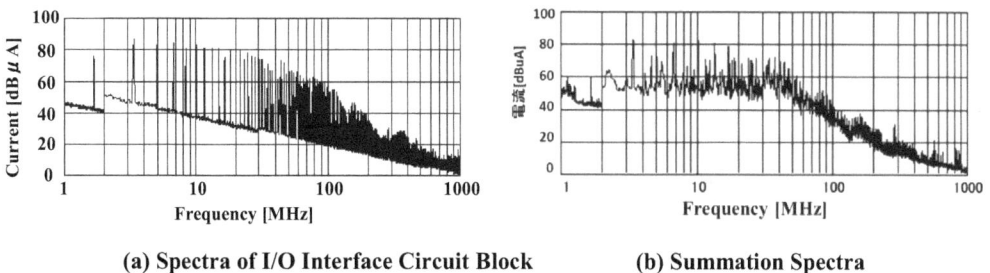

(a) Spectra of I/O Interface Circuit Block   (b) Summation Spectra

Figure 33: Measured Spectra of Power Current of 2nd Prototype of Mask ROM.

The driving performance reduces by reducing the voltage of the power supply. Fig. **34** shows the measured spectra of the power current of I/O interface circuit block of the original mask ROM and the second prototype of the mask ROM when the supply voltage changed from 4.6V to 5.5V. They were tested at the normal operation. The magnitude of the spectra of the power current of I/O interface circuit block of the first and second prototypes of the mask ROM were reduced almost 10dB when the supply voltage changed from 4.6V to 5.5V.

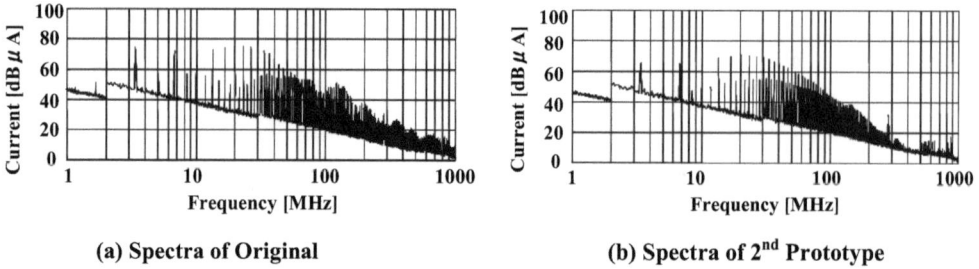

| (a) Spectra of Original | (b) Spectra of 2nd Prototype |

Figure 34: Measured Spectra of Power Current of I/O Interface Circuit Block of Mask ROM.

Fig. **35** shows the measured summation spectra of the power current of I/O interface and core circuit block of the original and the second prototype of the mask ROM when the voltage of the power supply was changed from 4.6V to 5.5V. The magnitude of the spectra of the power current of I/O interface circuit block of the first and second prototypes of the mask ROM were not changed when the supply voltage changed from 4.6V to 5.5V.

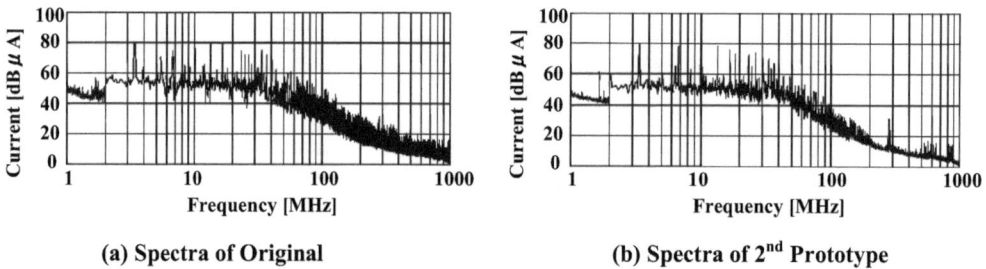

| (a) Spectra of Original | (b) Spectra of 2nd Prototype |

Figure 35: Measured Summation Spectra of Power Current of Mask ROM.

USB Controller

Fig. **36** shows the setup on the LSI tester for the measurement of the spectra of the power current of USB controller.

In Fig. **36**, the instruments are same as them of ASIC. The clock frequency of USB controller was 1.45MHz. The storage capacity was 8M byte. The voltage of the power supply was $5.0V$. Fig. **37** shows the spectra of the power current of the ambient noise on the power distribution line. Fig. **37a** shows the ambient noise at the standby of LSI tester, each Fig. **37a** and **37b** shows the ambient noise at the standby state and the operating state of LSI tester.

Figure 36: Setup for Measurement of Spectra of Power Current of USB Controller.

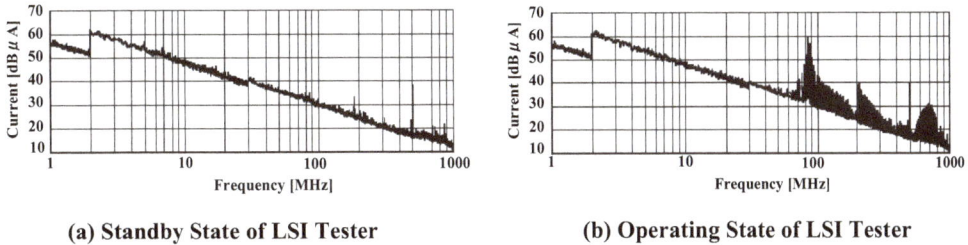

| (a) Standby State of LSI Tester | (b) Operating State of LSI Tester |

Figure 37: Measured Ambient Spectra of Power Current.

It was understood that the power consumption is reduced by shrinking the chip size of LSI.

Fig. **38** shows the measured spectra of the power current of the original and the first prototype of USB controller. The magnitude of the spectra of the power current of the original and the first prototype of USB controller is similar. From above, shrinking the chip size was considered to be not effective for suppressing EMI.

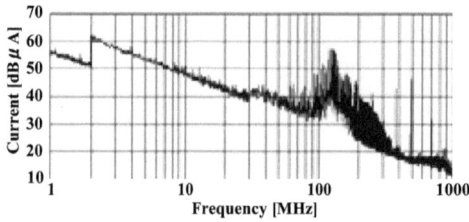

(a) Spectra of Original

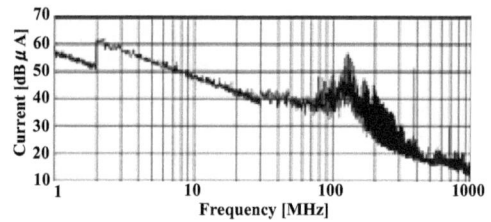

(b) Spectra of 1st Prototype

Figure 38: Measured Spectra of Power Current of USB Controller.

Gate Allay

The setup and the instruments were same as them of USB controller shown in Fig. **36**. Fig. **39** shows the measured spectra of the power current of the ambient noise on the power distribution line. Fig. **39a** shows the ambient noise at the standby of LSI tester, Each Fig. **39a** and **39b** shows the ambient noise at the standby state and the operating state of the LSI tester.

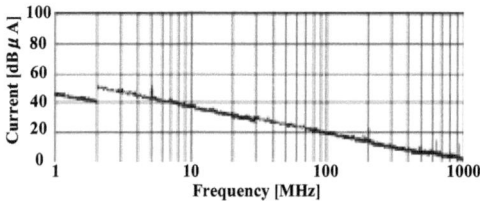

(a) Standby State of LSI Tester

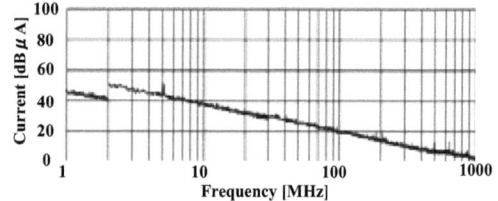

(b) Operating State of LSI Tester

Figure 39: Measured Ambient Spectra of Power Current.

It had been believed that EMI is suppressed by reducing the clock frequency and the supply voltage of LSI.

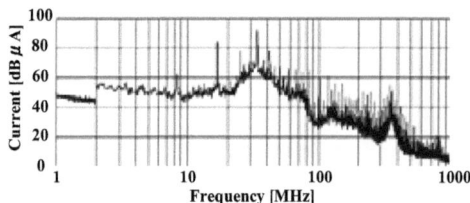

(a) 25MHz of Clock Frequency

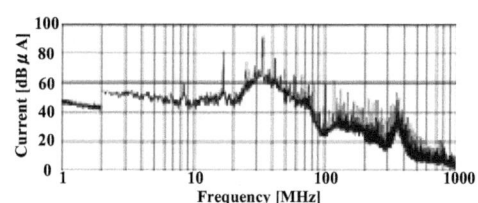

(b) 16.6MHz of Clock Frequency

Figure 40: Measured Spectra of Power Current of No.27 pin of Original Gate Allay.

Fig. **40** shows the measured spectra of the power current of No.27 pin of original Gate Allay when the clock frequency is changer to 16.6*MHz* from 25*MHz*. The magnitude of the spectra of the power current of No.27 pin of the original Gate Allay when the clock frequency is 16.6*MHz* is similar to it of 25*MHz*.

Fig. **41** shows the measured spectra of the power current of another No.150 pin of the original Gate Allay when the clock frequency is 16.6*MHz*. The magnitude of the spectra of the power current of No.150 pin of original Gate Allay is similar to it of No. 27 pin when the clock frequency is 16.6*MHz*.

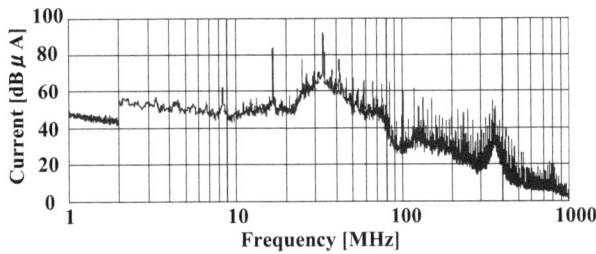

Figure 41: Measured Spectra of Power Current of No.150 pin of Original Gate Allay.

Fig. **42** shows the measured spectra of the power current of No.27 pin of original Gate Allay when the supply voltage is changed to 3.0*V* and 3.6*V* from 3.3*V*. The magnitude of the spectra of the power current of 3.0*V* and 3.6*V* is similar.

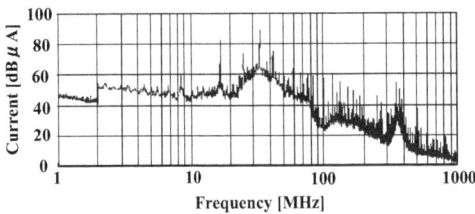

(a) 3.0V of Supply Voltage   (b) 3.6V of Supply Voltage

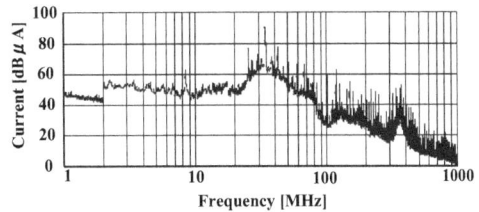

Figure 42: Measured Spectra of Power Current of No.27 pin of Original Gate Allay.

One-Chip Microcomputer

Fig. **43** shows the setup on the LSI tester for the measurement of the spectra of the power current of the microcomputer. The instruments are same as them of ASIC.

The voltage of the power supply was 4.6V. The clock frequency of on a chip was 55MHz and the supplied clock frequency was 6.6MHz.

Figure 43: Setup for Measurement of Spectra of Power Current of Microcomputer.

Fig. **44** shows the measured spectra of the power current. Fig. **44a** shows the spectra of the power current at the standby state of the tester and Fig. **44b** shows the operating state of the microcomputer.

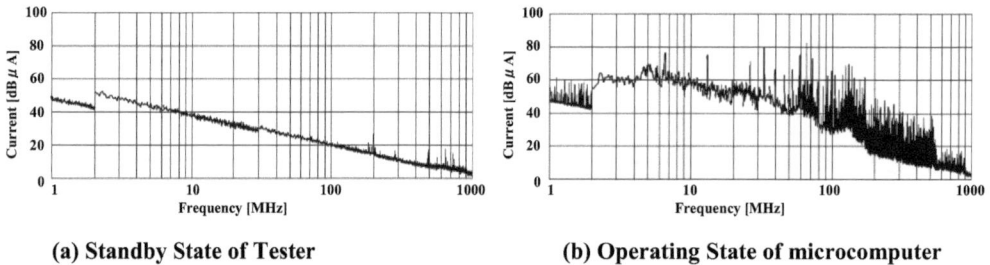

(a) Standby State of Tester   (b) Operating State of microcomputer

Figure 44: Measured Spectra of Power Current of Microcomputer.

SDRAM

Fig. **45** shows the setup on the LSI tester for the measurement of the spectra of the power current of the synchronous DRAM (SDRAM).

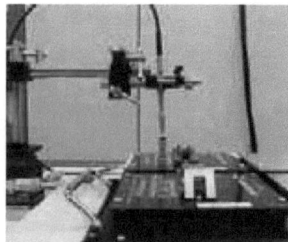

Figure 45: Setup for Measurement of Spectra of Power Current of SDRAM.

In Fig. **45**, the instruments are same as them of ASIC. The voltage of the power supply was $5.65V$. The clock frequency was $16.7MHz$ and $12.5MHz$. The storage capacity was 64 megabyte.

Fig. **46** shows the measured spectra of the power current on the power distribution line at the standby state of the memory tester.

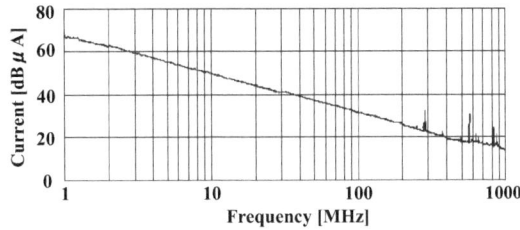

Figure 46: Measured Power Current of Memory Tester at Standby State.

Fig. **47** shows the measured spectra of the power current on the power distribution line at the standby state of SDRAM when the 1st program is running on the memory tester. Each Fig. **47a** and **47b** shows the spectra of the power current at operating by the clock of $16.7MHz$ and $125MHz$.

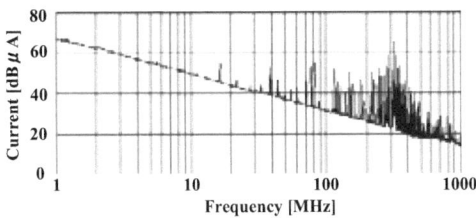

(a) Spectra at operating by 16.7MHz Clock   (b) Spectra at operating by 125MHz Clock

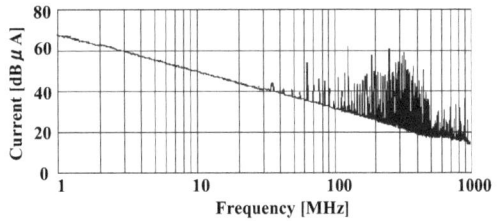

Figure 47: Measured Spectra of Power Current of SDRAM at Standby of 1st Program.

It had been believed that EMI is suppressed by reducing the clock frequency and the supply voltage of LSI.

Fig. **48** shows the measured spectra of the power current of SDRAM when the clock frequency is changed to $16.7MHz$ from $125MHz$. The magnitude of the spectra of the power current of SDRAM increased when the clock frequency is changed to $16.7MHz$ from $125MHz$.

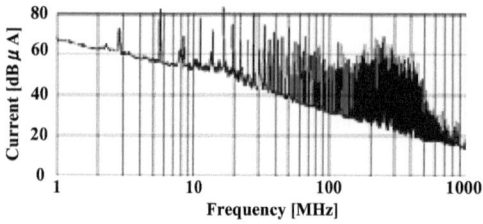

(a) 16.7MHz Clock   (b) 125MHz Clock

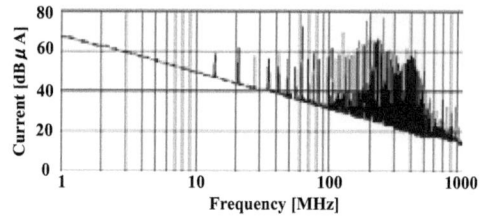

Figure 48: Measured Spectra of Power Current of SDRAM at Operating of 1st Program.

Fig. **49** shows the measured spectra of the power current on the power supply line at each standby state and operating state when the second program is running on the memory tester at clock of 125*MHz*.

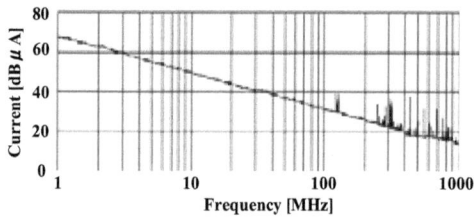

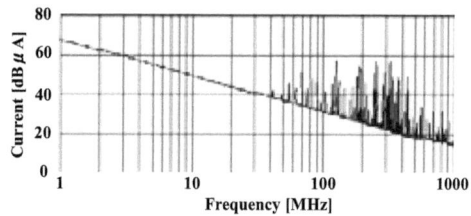

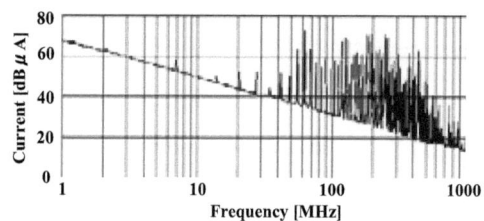

(a) Standby State   (b) Operating State

Figure 49: Measured Spectra of Power Current of SDRAM when 2nd Program is Running.

Fig. **50** shows the measured spectra of the power current of SDRAM at each standby state and operating state when the third program is running on the memory tester at clock of 125*MHz*.

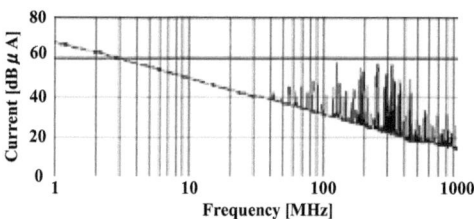

(a) Standby State   (b) Operating State

Figure 50: Measured Spectra of Power Current of SDRAM when Third Program is Running.

From Fig. **48**, Fig. **49**, and Fig. **50**, it was clarified that the magnitude of the spectra of the power current of SDRAM is inclined to be influenced by the test program and magnitude of the spectra of the power current of SDRAM is not necessarily increased when the clock frequency increases to $125MHz$ from $16.7MHz$.

From above, the measurement of the spectra of the power current of LSI by using the MP method is supposed to be effective for the study and design of suppressing EMI. It was confirmed that the electric field strength radiation of the test board depends on the magnitude of the spectra of the power current.

FEASIBILITY OF SUPPRESSING EMI OF LSI

From above EMI evaluation, it was clarified that the electric field radiation of the test board depends on the magnitude of the spectra of the power current largely. The improvement of the power decoupling circuit is effective for reducing the magnitude of the spectra of the power current.

Fig. **51** shows an example of the on-chip decoupling circuit which was formed on the prototype of the 32bit RISC processor. Fig. **51a** shows an example of the 1ˢᵗ cell of the on-chip capacitor, Fig. **51b** shows an example of the layout of the on-chip decoupling circuit, and Fig. **51c** shows an example of the 2ⁿᵈ cell of the on-chip capacitor.

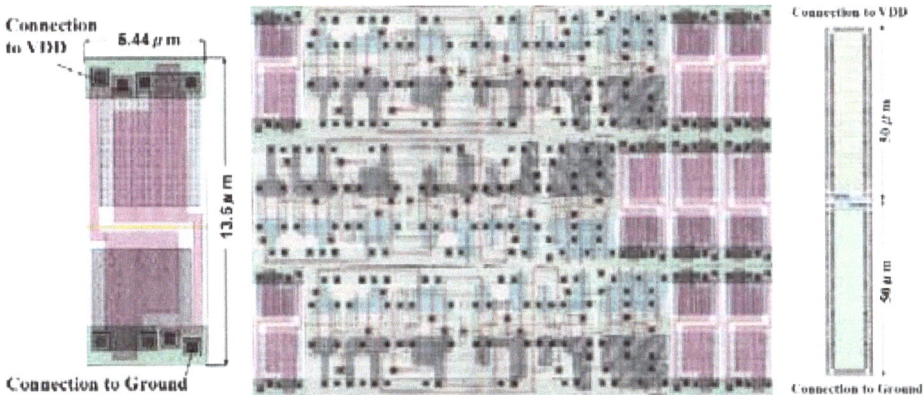

(a) 1ˢᵗ **Capacitor Cell**   (b) **Layout of On-chip Decoupling Circuit** (c) 2ⁿᵈ **Capacitor Cell**

Figure 51: On-chip Decoupling Circuit Formed on Prototype of 32bit RISC Processor.

In Fig. **51**, the 1ˢᵗ cell of the on-chip capacitor having the capacitance of $0.11pF$ and the resistance of 1Ω was formed by the extended gate capacitor of the on-chip inverter in the 4 grids, In Fig. **51b**, the on-chip decoupling circuit at the transistor area was formed by replacing some inverters to the 1st capacitor cells which have summation capacitance of 5.6nF. In Fig. **51c**, the 2nd cell of the on-chip capacitor having the capacitance of 5.1pF and the resistance of 0.2Ω was formed by the metals and the insulator on the power supply wiring layers and the summation capacitance on the power supply wiring layer was 4.0nF. As the result, summation capacitance of the added capacitor on the prototype of 32bit RISC processor was 7.6nF.

Fig. **52** shows the measured spectra of the power current of the core circuit of the original and the prototype of 32bit RISC processor. The magnitude of the spectra of the power current of the prototype is smaller than it of the original up to 200MHz approximately however the decoupling effect was not confirmed at over 400MHz.

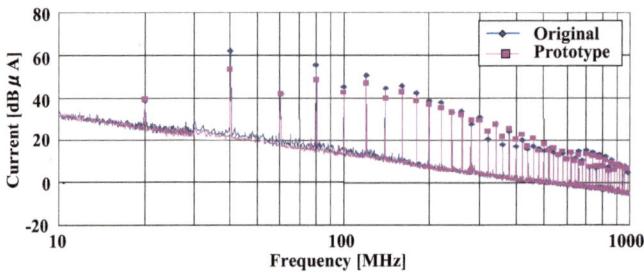

Figure 52: Measured Spectra of Power Current of Core Circuit of 32bit RISC Processor.

Fig. **53** shows the snap of the measurement of the spectra of the power current of the microprocessor.

Figure 53: Setup for Measurement of Spectra of Power Current of Microprocessor.

In Fig. **53**, the microprocessor which consisted of the macro of USB Hub and CPU was manufactured for the key board of the computer, and the improved magnetic probe was used for the measurement of the power current. This microprocessor was prototyped by shrinking the flip-flop cell and adding the 1st and 2nd capacitor cells. The power supply current was reduced by shrinking the flip-flop cell. The summation capacitance of each 1st and 2nd capacitor cells of USB Hub macro was 950pF and 620pF, and the summation capacitance of each 1st and 2nd capacitor cells of CPU macro was 260pF and 590pF.

Fig. **54** shows the measured spectra of the power current of the core circuit of the original and the prototype of the microprocessor. The magnitude of the spectra of the power current of the prototype was reduced about 10dB against the original of the microprocessor in a broad frequency range.

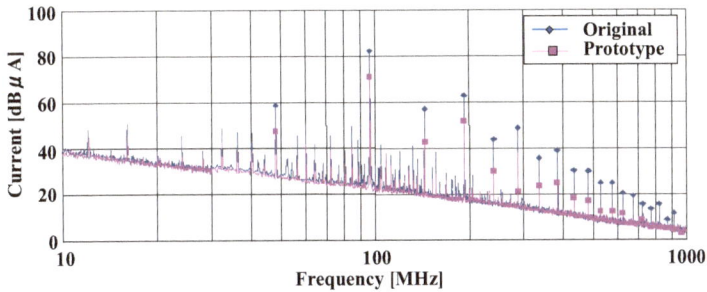

Figure 54: Measured Spectra of Power Current of Microprocessor.

Fig. **55** shows the measured radiated EF strength of the ley board which is used the microprocessor.

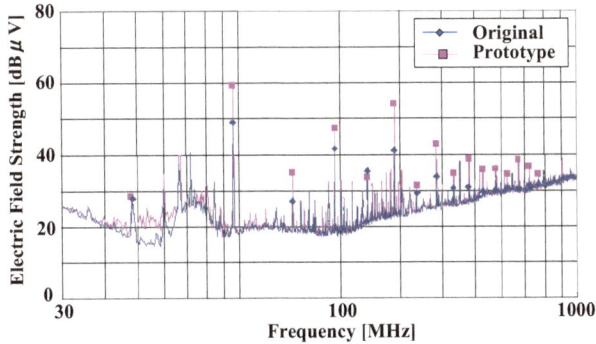

Figure 55: Measured Radiated EF Strangth of Key Board used Microprocessor.

In Fig. **55**, the magnitude of the radiated EF strength of the key board was reduced about 10dB against the original of the microprocessor in a broad frequency range. The tendency of the reduction of the radiated EF strength was well matched to it of the spectra of the power current.

Fig. **56** shows the chip formations of the prototype of the gate allay. The chip size of the original gate allay was 5.5*mm* square, the gate allay was prototyped by expanding the chip size to 6.54*mm* square and adding the capacitor cells having the summation capacitance of 20.7*nF*, and the all pink cells were the capacitor cells. Fig. **56** is considered to be being shown the physical limit of the improvement by the on-chip decoupling technique.

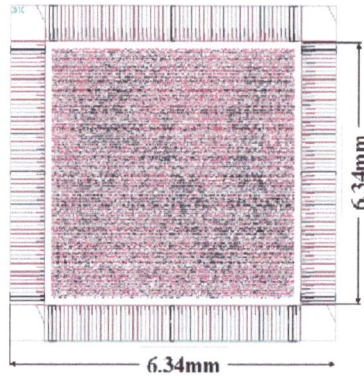

Figure 56: Chip Formations of Prototype of Gate Allay.

Fig. **57** shows the measured spectra of the power current of the core circuit of the original and the prototype of the gate allay. The magnitude of the spectra of the power current of the prototype was reduced about 15dB against the original of the microprocessor in a broad frequency range.

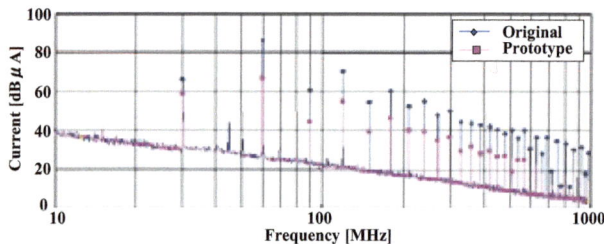

Figure 57: Measured Spectra of Power Current of Prototype Gate Allay.

Fig. **58** shows an example of the measured spectra of the signal current of the original and the prototype of the gate allay. Each Fig. **58a** and Fig. **58b** shows the spectra at each terminal of the logic 1 and it of the logic 0.

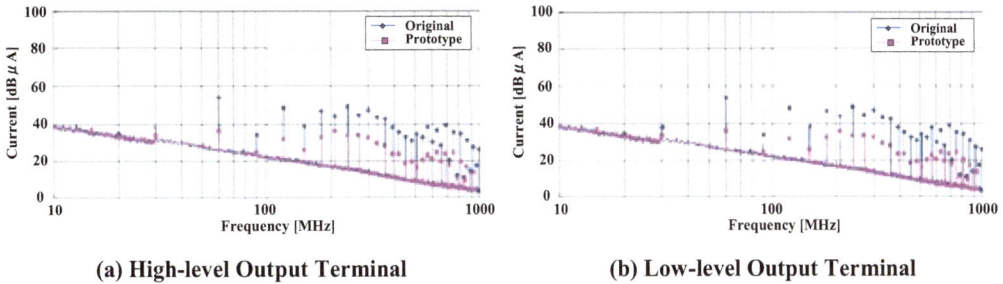

(a) High-level Output Terminal   **(b) Low-level Output Terminal**

Figure 58: Measured Spectra of Signal Current of Original and Prototype of Gate Allay.

In Fig. **58**, the shape of the spectra of the signal current of the logic 0 and it of the logic 1 is similar. And the magnitude of the spectra of the signal current at the terminals of the logic 1 and it of the logic 0 of the prototype are 10-50dB smaller than the spectra of the power current up to 200MHz but the shape of the spectra at the terminals of the logic 1 and it of the logic 0 of the original prototype are similar at over 300MHz.

From above, it was clarified that the effect of EMI suppressing of the on-chip capacitors for the power decoupling circuit is lower than 15dB. Therefore, the relatively large amount of EMW is considered to be leaked to the signal line from PDN of LSI. It was also confirmed that the shape of the spectra of the signal current is similar to it of the spectra of the power current over than 50MHz.

REFERENCES

[1]   http://www.nec.co.jp/solution/engsl/pro/emc_scanner/pdf/emc_catalog.pdf
[2]   http://www.iec.ch/dyn/www/f?p=103:91:0::::FSP_LANG_ID:25#q=IEC 61967

Send Orders for Reprints to reprints@benthamscience.net

CHAPTER 6

Feasibility of Reconfiguring SMC to QSCC

Abstract: EMC problems will be solved if SMC is formed by the quasi stationary state closed circuit (QSCC). The result of the feasibility study for reconfiguring SMC to QSCC is presented.

Keywords: Quasi-stationary, Quasi stationary state closed circuit (QSCC), signal line, microstrip line, Fourier Transform, signal delay, signal attenuation, crosstalk, mutual capacitance, electromagnetic emission, signal reflection, signal dispersion, clock frequency, FPGA, EF strength, limit length, distribution of MF, CAD.

LIMITATION OF SIGNAL LINE LENGTH ABOUT RADIATED EMISSION

If all electromagnetic fields are quasi-stationary or if they are slowly varying, according to the electromagnetism, the displacement current or EMW neglected. Therefore, the EMI does not occur on the QSCC. QSCC can be formed by limiting the wire length according to the wavelength of EMW on PCB. The magnitude of EMWs on PDN is relatively large because PDN on the chip and the board is formed by the continuous metal wires and traces. In addition, EMW on PDN have the broad spectrum. EMWs arrive to PDN from LSI or the transistor and the power supply line. When the slender power traces are formed and the I/O signal traces are separated from the clock traces, EMI of the PDN will be suppressed effectively and the magnitude of EMWs on the signal line will reduced be reduced by shortening of the signal line. The circuit situation such as the signal delay, the signal attenuation, the cross-talk, the electromagnetic emission, the signal reflection, and the signal dispersion will be also improved greatly when SMC is reconfigured to QSCC.

Fig. **1** shows an example of the test board for getting the parameters of the design rule.

In Fig. **1**, the test board was formed by FR4, the line length of the signal line was 100*mm*, FPGA was EPF10K100E(240) of ALTERA, ROM was EPC2(20) of ALTERA, the frequency of the ring oscillator was 70*MHz*, the duty cycle was 50%, the rise time of the signal voltage was 1*ns*, and the capacitance of the load capacitor

Hirokazu Tohya

was $5pF$, the microstrip Line in Fig. **1** consisted of the strip conductor which has each thickness and width of $18\mu m$ and $180\mu m$, the insulator of which thickness is $285\mu m$, the chip-size and the lead-frame of FPGA are considered to be $12mm$ and $15mm$, and the ring oscillator of $70MHz$ was formed by connecting thirty seven primitive cells in series on FPGA.

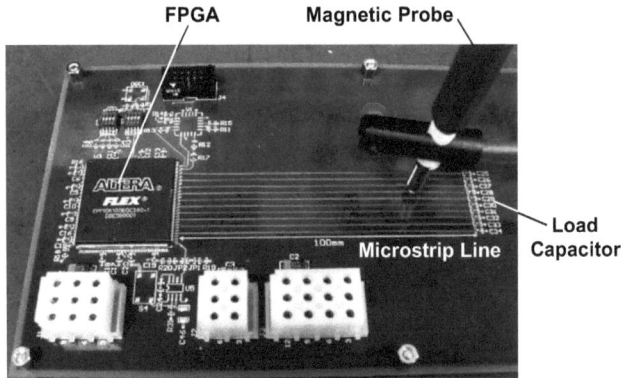

Figure 1: Example of Test board.

Fig. **2** shows the circuit diagram of the signal line on the test board. Fig. **2a** shows the circuit diagram and Fig. **2b** shows the cross section of the microstrip line as the signal line. The dielectric constant of the insulator was 4.35.

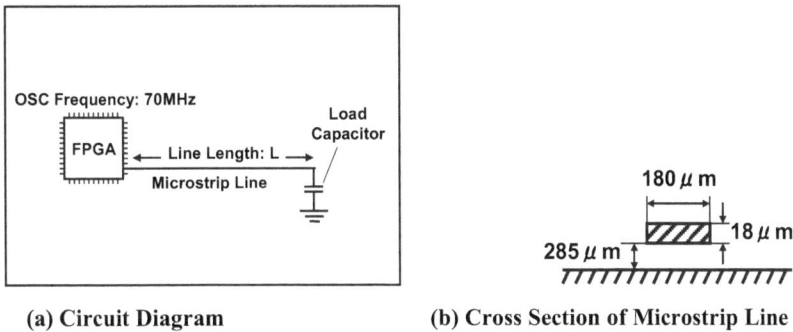

(a) **Circuit Diagram**   (b) **Cross Section of Microstrip Line**

Figure 2: Circuit Diagram of Signal Line on Test Board.

Fig. **3** shows the voltage shape and the calculated spectra on the signal line of the test board. Fig. **3a** is the voltage shape. Fig. **3b** shows the calculated spectra of the signal voltage by the Fourier Transform. In Fig. **3a**, T was $14.08ns$, τ was $7.04ns$, τ_r was $1.7ns$ τ_f was $0.5ns$, and the level A was $2.5V$. In Fig. **3b**, the almost all spectra exist up to 1*GHz*. The wave length of 1*GHz* on the signal line is 14.4*mm*. Therefore, the lead-frame of FPGA and the signal line on the board should be designed for the QSCC.

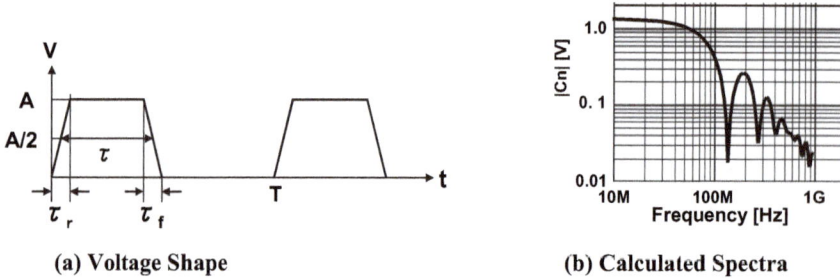

(a) Voltage Shape   (b) Calculated Spectra

Figure 3: Voltage Shape and Calculated Spectra on signal line of the test board.

Fig. **4** shows the measured strength of EF at 3*m* distance when the length of the signal line on the board is 35*mm*, 15*mm*, 1*mm*, and no signal.

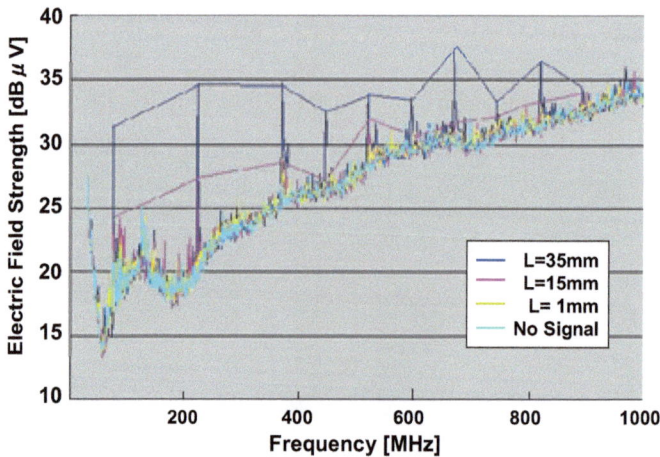

Figure 4: Measured Strength of EF at 3*m* Distance of the test board.

Fig. **5** shows the spectra of the test board when the length of the signal line is 35*mm*. Fig. **5a** shows the measured current of the signal line. Fig. **5b** shows the calculated and measured spectra of the EF strength at 3*m* distance of the test board. In Fig. **5b**, the spectra of the EF strength were calculated by the equation 35 of Glossaries. In Fig. **5a**, the EF strength of E (I_1) and E (I_2) was calculated by each 0.5% of I_1 and 5% of I_2. Each percent is assumed CM radiation.

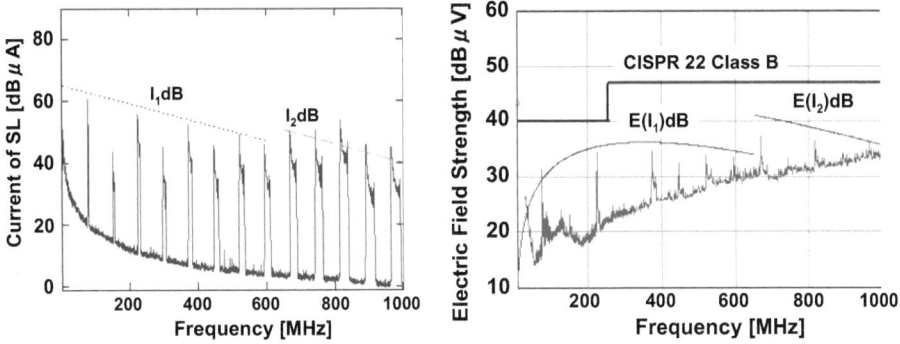

(a) Measured Current of Signal Line   **(b) Calculated and Measured EF Strength**

Figure 5: Spectra of of Test Board of 35*mm* Length of Signal Line.

Fig. **6** shows the measured current of the signal line and the calculated and measured spectra of the EF strength at 3*m* distance of the test board when the length of the signal line is 15*mm*. In Fig. **6a**, the calculated spectra of the EF strength were got by by the equation 35 of Glossaries and the calculation condition is same as the line length of 35*mm*.

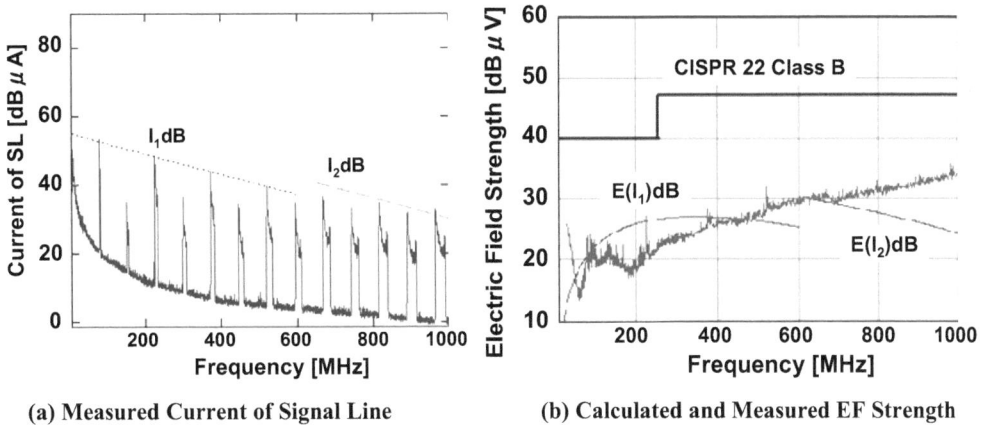

(a) Measured Current of Signal Line   **(b) Calculated and Measured EF Strength**

Figure 6: Spectra of EF Strength of Test Board of 15*mm* Length of Signal Line.

In Fig. **6b** and Fig. **6b**, the calculation condition was considered to be reasonable for setting the limit of the signal line for forming QSCC.

Fig. **7** show the limit length corresponding to the clock frequency when the FPGA of the test board is used.

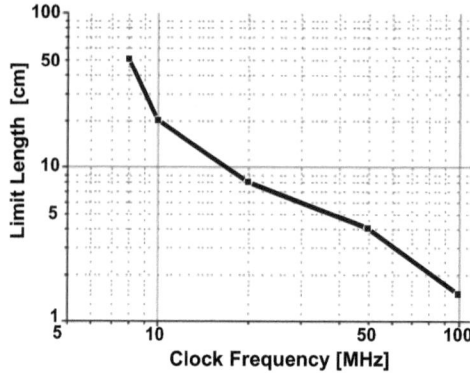

Figure 7: Limit Length Corresponding to Clock Frequency.

DISCUSSION OF SIGNAL LINE LENGTH ABOUT CROSSTALK

The signal on the transmission line tends to be coupled to the other transmission line electro-magnetically. This phenomenon is called as the crosstalk which is the most difficult problem when the length of the signal trace is relatively long [1, 2]. The cross talk voltage is proportional to the magnitude of the transient current and the transient voltage.

The cross talk voltage which is excited by the current on the transmission line is:

$$E_C = \int_{X_1}^{X_2} L_m \frac{dI(x)}{dt} dx \tag{1}$$

where E_C is the inducted voltage on the transmission line, L_m is the mutual inductance between the transmission lines, I is the source current on the transmission line, and each X_1 and X_2 are the terminals of the parallel transmission line.

The cross talk current which is excited by the voltage on the signal line is:

$$I_C = \int_{X_1}^{X_2} C_m \frac{dV(x)}{dt} dx \tag{2}$$

where I_C is the inducted current on the transmission line, C_m is the mutual capacitance between the transmission lines, and V is the voltage on first transmission line.

As mentioned in Chapter 2, the EF is spreading from the microstrip line 1, and is influencing to the microstrip line 2 largely. The influenced EMW causes the crosstalk problem. In Fig. **15** of the chapter 2, the characteristic curve of S_{21} of the microstrip line 1 is excellent because this magnitude is almost 0dB, the characteristic curve of S_{11} of the microstrip line 1 is excellent because this value is almost -75dB (=5.6×10-4), although, the characteristic curve of S_{41} of the microstrip line 2 or the crosstalk is not good because this magnitude exists between -15dB (=18%) to -10dB (=52%).

Fig. **8** shows the circuit diagram of the test board for measurement of the cross talk. L_X was 60*mm*, and the distance between the strip conductors of the 1st microstrip line and the 2nd microstrip line was 100*μm*.

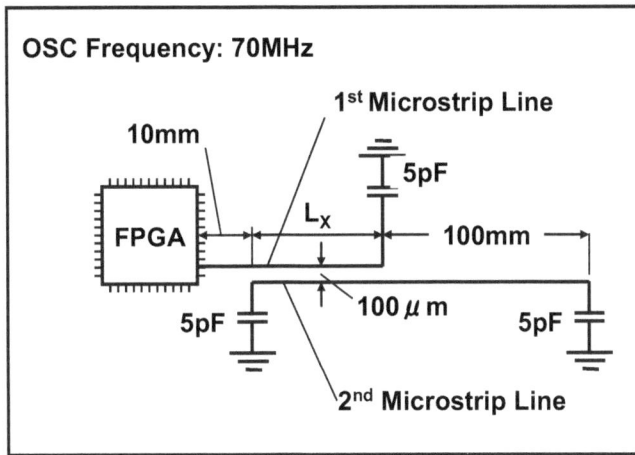

Figure 8: Circuit Diagram of Test Board for Measurement of Cross Talk.

Fig. **9** shows the voltage shape on the signal line. Fig. **9a** shows the measured voltage shape of the signal and the crosstalk. Fig. **9b** shows the calculated voltage shape of the crosstalk. The maximum positive magnitude of the cross talk voltage was 1*V* which is smaller than the half voltage of the signal. In Fig. **9b**, C_m in the equation 1 was got by the simulator APSIM RLGC and the waveform was simulated by APSIM Spice.

Fig. **10** shows the measured spectra of the signal and the cross talk of the test board.

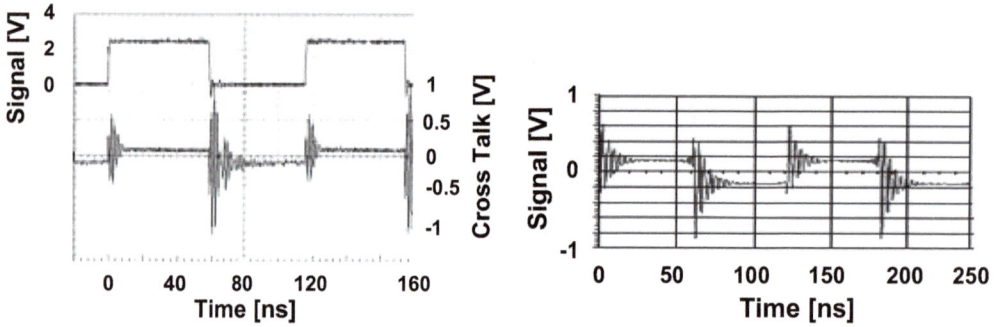

(a) Measured Voltage Shape of Signal and Crosstalk   (b) Calculated Voltage shape Crosstalk

Figure 9: Voltage Shape on Signal Line.

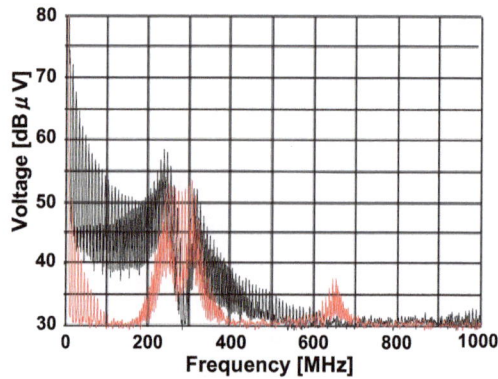

Figure 10: Measured Spectra of Signal and Cross Talk of Test Board.

From above discussion, above mentioned measurement condition was considered to be suitable for forming QSCC.

DESIGN AND PROTOTYPING OF QSCC TEST BOARD

Fig. **11** shows the prototyped test board. Ten FPGA were used. The design condition was the following; the size is $300mm \times 300mm \times 1.6mm$, the substrate is made of FR4, six layers which consist of the first layer of the signal trace, the second layer of the ground plane, the third layer of the $2.5V$ power plane which is used for core block of FPGA, the fourth layer of the $3.3V$ power plane which is used for the I/O block of FPGA and the other devices, the fifth layer of the ground plane, and the sixth layer of the signal trace, thickness of the first and sixth copper layers is $18\mu m$, thickness of the other copper layers is $35\mu m$, the pitch of the signal trace is

0.36mm, the width of the signal trace is 0.18mm, the minimum diameter of the *via* is 0.3mm, and the minimum diameter of the via-pad is 0.6mm.

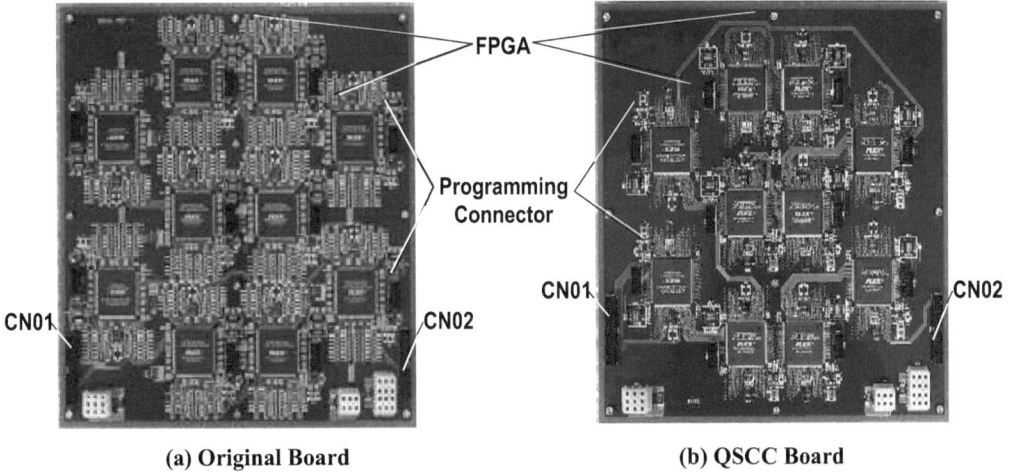

(a) Original Board   (b) QSCC Board

Figure 11: Prototyped Test Board.

Each ring oscillator of 15MHz, 30MHz, and 60MHz was formed by connecting eight and sixteen and thirty two primitive cells in series on FPGA as the signal. Each FPGA had eighty signal drivers and eighty signal receivers. The fixed-frequency signals were distributed reciprocally by eighty signal traces from the driver to the receiver of ten pieces of FPGA by the eight hundred signal traces. The random-frequency signals were also formed by the random signal generator which consists of forty-bit linear feedback shift resistor and they were supplied by the other drivers on FPGA. The random pattern of 1012 was generated by 166 minutes operation when the clock frequency is 100MHz. The different four kinds of voltage were supplied to the test board by the 12V lead batteries through the four kind of the linear power supply unit. The three types of the fixed-frequency signals and the random-frequency signals on the signal lines between FPGA were set through each programming connector of FPGA and the typical signals were monitored by CN01 and CN02. The original board was designed by the commercialized CAD system and in accordance with the current design rule and QSCC board was designed by the commercialized CAD system and in accordance with the above mentioned wiring rule.

Fig. **12** shows the circuit diagram of QSCC Test Board.

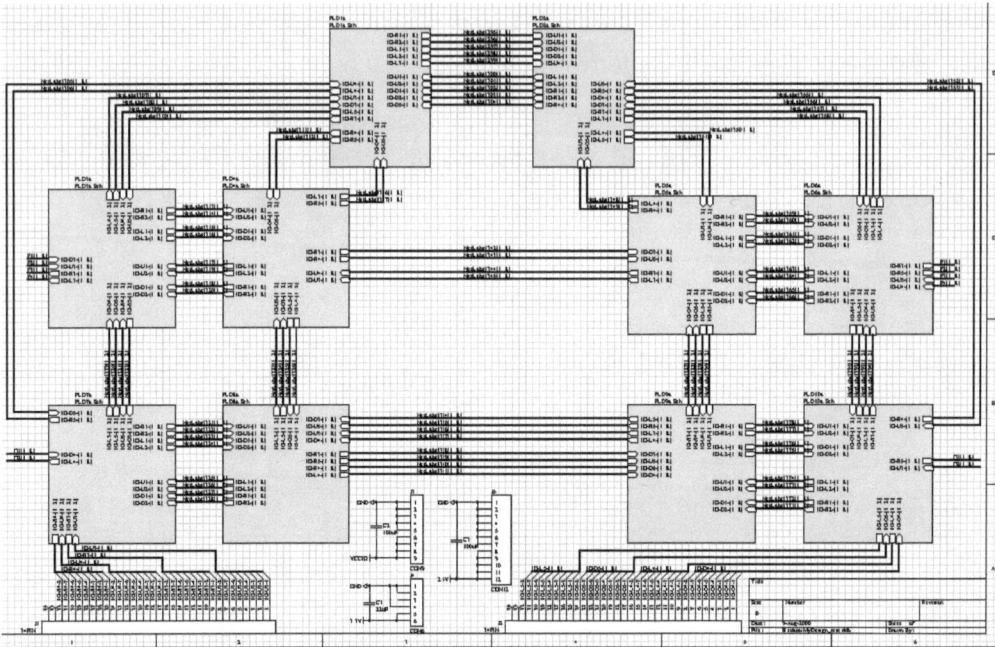

Figure 12: Circuit Diagram of QSCC Test Board.

Fig. **13** show the first signal distribution of the QSCC test board.

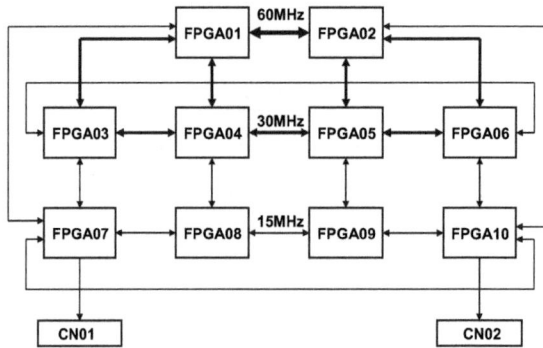

Figure 13: First Signal Distribution of QSCC Test Board.

In Fig. **13**, forty input signals and forty output signals were distributed between all FPGA. The fixed-frequency signals of 60*MHz* were distributed only between FPGA01 and FPGA02. The fixed-frequency signals of 30*MHz* were distributed between FPGA01 and FPGA03, between FPGA03 and FPGA04, between FPGA04 and FPGA05, between FPGA05 and FPGA06, and between FPGA06 and FPGA02. The fixed-frequency signals of 15*MHz* were distributed between FPGA01 and

FPGA07, between FPGA07 and FPGA08, between FPGA08 and FPGA09, between FPGA09 and FPGA10, between FPGA10 and FPGA07, between FPGA10 and FPGA02, between FPGA03 and FPGA06, from FPGA07 to CN01, and from FPGA10 to CN02. These wiring route were not same between the original board and QSCC board.

Fig. **14** show the second signal distribution of the QSCC test board. Each two fixed-frequency signals of 15*MHz* were distributed interactively between FPGA01 and FPGA02, between FPGA01 and FPGA03, between FPGA02 and FPGA06, and between FPGA08 and FPGA09. These signal lines were formed on only surface of the PCB, and the number of the vias was two. The number of the signal lines was sixty one. Each length of the signal lines was different but each wiring route was same between the original board and QSCC board.

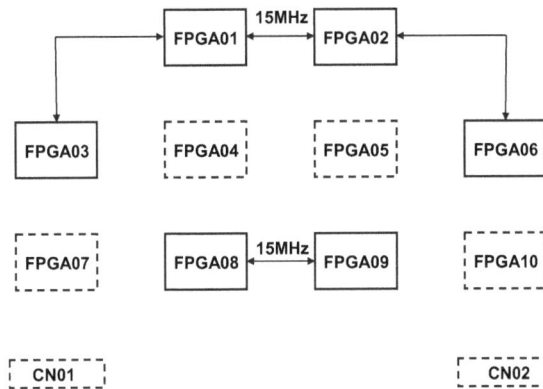

Figure 14: Second Signal Distribution of QSCC Test Board.

Fig. **15** show the third signal distribution of the QSCC test board.

In Fig. **15**, forty input signals and forty output signals were distributed between all FPGA, seven fixed-frequency signals of 60*MHz* were distributed between FPGA01 and FPGA02, each eight fixed-frequency signals of 30*MHz* were distributed between FPGA01 and FPGA03 and between FPGA03 and FPGA04, nine fixed-frequency signals of 30*MHz* were distributed between FPGA02 and FPGA06, twenty four fixed-frequency signals of 15*MHz* were distributed between FPGA08 and FPGA09, and seven hundred thirty nine random-frequency signals which consist of the pulse signal of 80*ns* width were distributed with all other signal lines between each FPGA. The actual IT board was simulated by the third signal distribution.

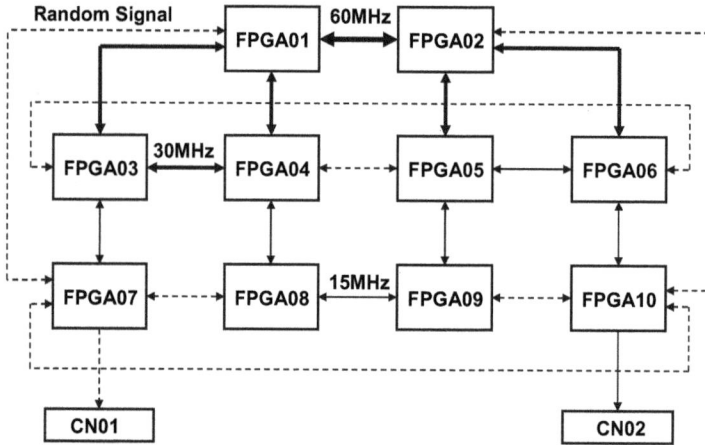

Figure 15: Third Signal Distribution of QSCC Test Board.

EVALUATION RESULT

Fig. **16** show the measured strength of EF of the first signal distribution of the original board. The strength of EF was measured at $3m$ distance, some spectra of the EF were larger than the limit of IEC/CISPR22 Class B, and one spectrum of the EF at the horizontal polarization was 14dB larger than the limit.

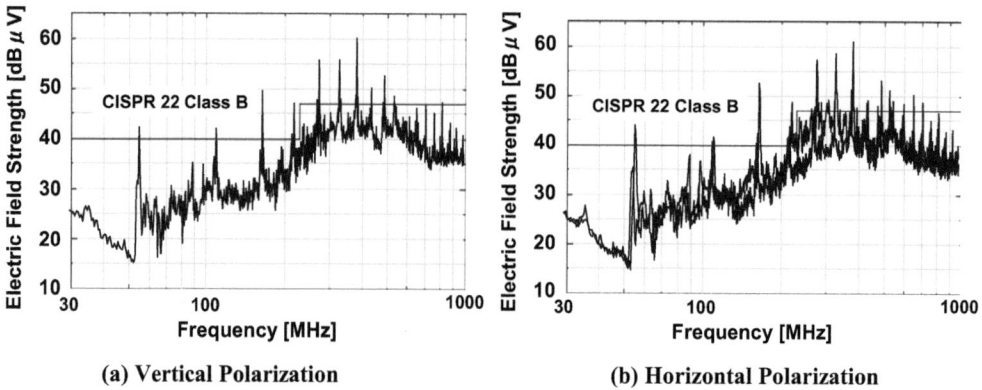

(a) Vertical Polarization   (b) Horizontal Polarization

Figure 16: Measured Strength of EF of First Signal Distribution of Original Board.

Fig. **17** show the measured strength EF of of the first signal distribution of QSCC board.

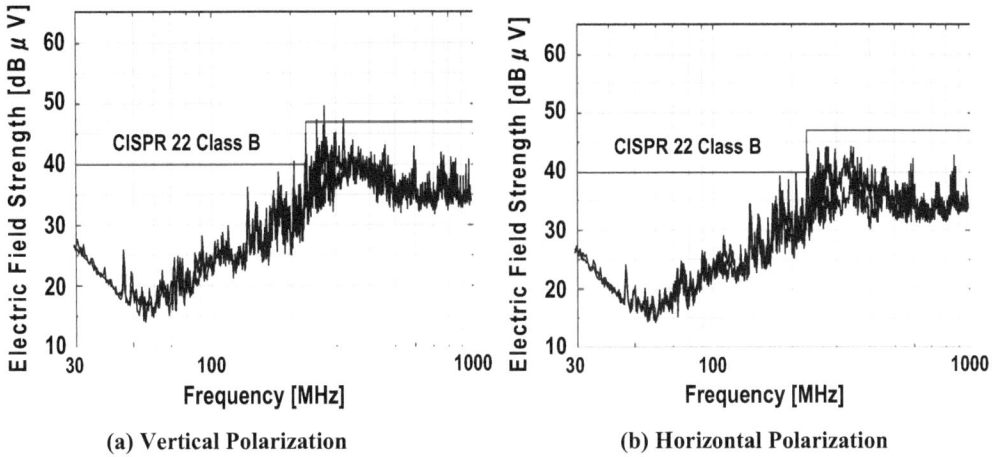

(a) Vertical Polarization   (b) Horizontal Polarization

Figure 17: Measured Strength of EF of First Signal Distribution QSCC Board.

In Fig. **17**, one spectrum of the EF at the vertical polarization was about 3dB larger than the limit of IEC/CISPR22 Class B, and the spectra of EF at the horizontal polarization were 3dB smaller than the limit of IEC/CISPR22 Class B.

Fig. **18** show the measured strength of EF of the second signal distribution of the original board. The spectra of EF at the horizontal polarization have met the Class B of CISPR22 barely.

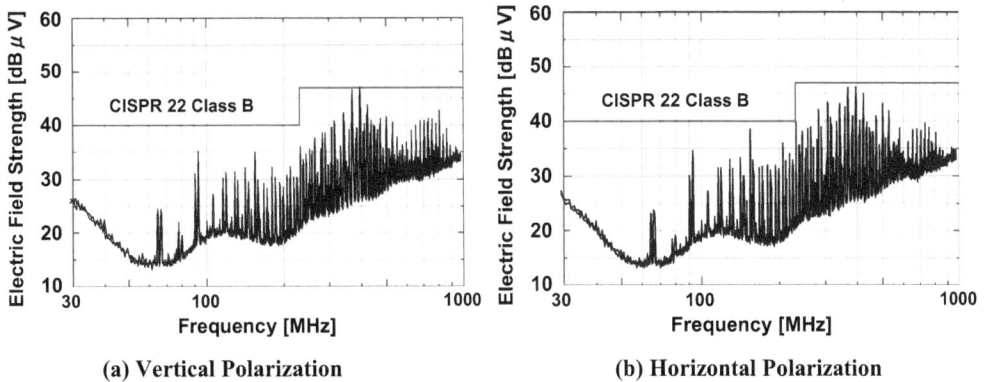

(a) Vertical Polarization   (b) Horizontal Polarization

Figure 18: Measured Strength of EF of Second Signal Distribution of Original Board.

Fig. **19** show the measured strength of EF of the second signal distribution of QSCC board. The spectra of the EF at the horizontal polarization have met the class B of CISPR22 with the margin of 11dB.

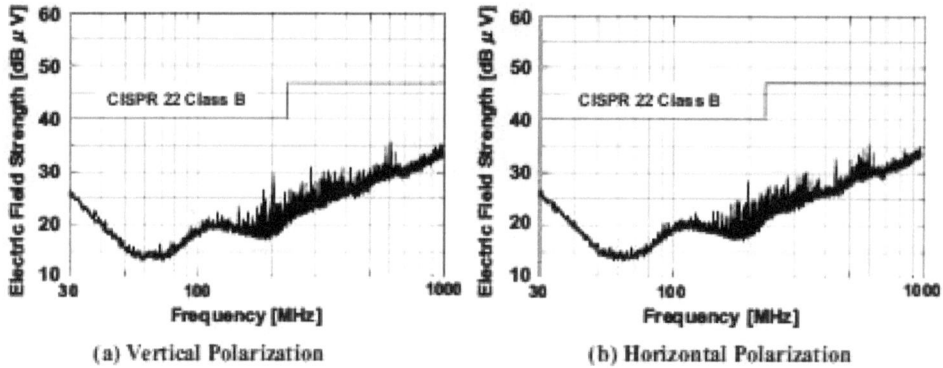

(a) Vertical Polarization   (b) Horizontal Polarization

Figure 19: Measured Strength of EF of Second Signal Distribution of QSCC Board.

Fig. 20 show the measured strength of EF of the third signal distribution of the original board.

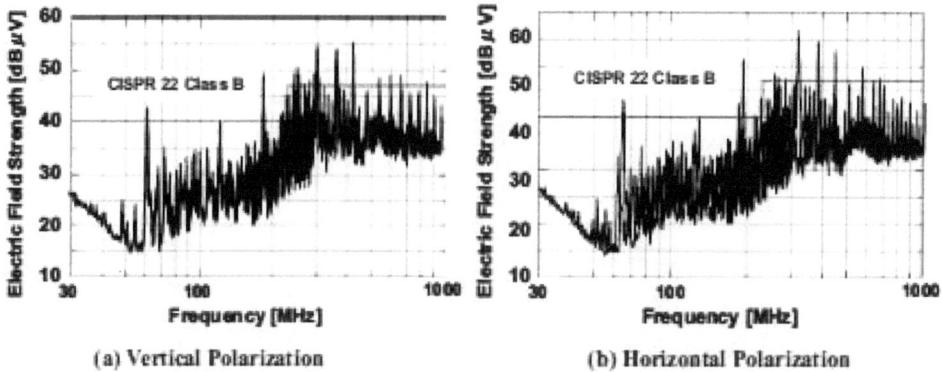

(a) Vertical Polarization   (b) Horizontal Polarization

Figure 20: Measured Strength of EF of Third Signal Distribution of Original Board.

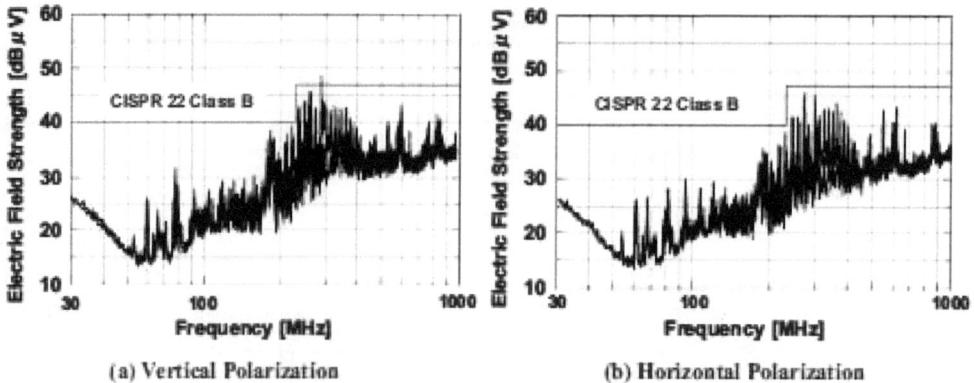

(a) Vertical Polarization   (b) Horizontal Polarization

Figure 21: Measured Strength of EF of Third Signal Distribution of QSCC Board.

Fig. **21** show the measured strength of EF of the third signal distribution of QSCC board. One spectrum of the EF strength at the vertical polarization was about 2dB larger than the limit of IEC/CISPR22 Class B, and the spectra of the EF strength at the horizontal polarization were 1dB smaller than the limit of IEC/CISPR22 Class B.

Fig. **22** show the measured distribution of MF of the back side of the third signal distribution board.

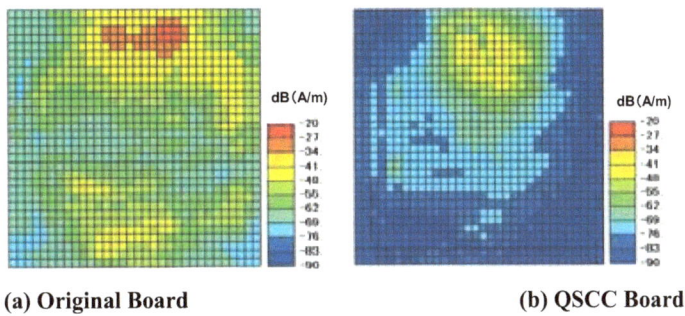

(a) Original Board   (b) QSCC Board

Figure 22: Measured Distribution of MF of Back Side of 3rd Signal Distribution Board

CONCLUSIONS

It was confirmed that the radiated emission was improved 10dB approximately when the third signal distribution was designed as the QSCC. The test condition of the radiated emission of the test board of the third signal distribution is similar to it of the standalone MB shown in Fig. **14a** in Chapter **4**. The test result was that the prototype MB is improved 10dB approximately against the original MB. Therefore, it was considered that the improvement of the radiated emission of three test boards depends on the decoupling circuit of the PDN.

REFERENCES

[1]   J. Zhang, E.G. Friedman, "Effect of shield insertion on reducing crosstalk noise between coupled interconnects" IEEE Circuits and Systems, ISCAS '04, vol. 2, pp. 529-532, 2004.

[2]   S. Delmas-Bendhia, F. Caignet, E. Sicard, M. Roca, "On-Chip Sampling in CMOS Integrated Circuits" IEEE Circuits Systems, IEEE TRANSACTIONS ON ELECTROMAGNETIC COMPATIBILITY, vol. 41, No. 4, pp. 403 -406, 1999.

Send Orders for Reprints to reprints@benthamscience.net

CHAPTER 7

Feasibility of Suitable Decoupling Component for QSCC

Abstract: The improvement of the decoupling performance of PDN will have the highest priority for forming QSCC. The result of the feasibility study of the suitable decoupling component is presented.

Keywords: Low-impedance line structure component (LILC), QSCC, EMI, PDN, EMW, AC circuit, terminal impedance, transmission coefficient, chip ceramic LILC, multi-layers chip ceramic LILC, strip line, Pb (Zr, Ti) 03 system, loss tanδ, ceramic cylinder LILC, FPGA, solid aluminum LILC, carbon paste, conductive polymer, carbon graphite, etched aluminum film.

CHIP CERAMIC LILC

It was confirmed that the idea of QSCC is effective for reducing EMI but the performance of the conventional decoupling circuit for PDN is limiting the suppression of the radiated emission. Therefore the review of the characteristics of capacitor at the high-frequency band became necessary. And the novel decoupling component which can work to EMW effectively was discussed. The AC circuit including the switching mode circuit is EMW circuit because the electric field or the magnetic field becomes EMW when it changes in accordance with the electromagnetism. Therefore, the transmission line technologies became necessary for improvement of the performance of the decoupling circuit. At first, the formation method in PCB was discussed. The terminal impedance and the transmission coefficient will be small enough when the thickness of the insulator layer is thin enough and the dielectric constant of the insulator is large enough even if the slender traces are used to PDN. Unfortunately, it was confirmed by the prototyping that this method is quite difficult for manufacturing. Therefore, the study of the low-impedance line structure component (LILC) which consists of the proven manufacturing method and materials of the capacitor was started. First the chip ceramic LILC was discussed because the chip ceramic capacitor has been considered to be suitable for the high-frequency application and has been being used to the high-performance decoupling circuit.

Fig. **1** shows the prototype of the chip ceramic LILC of the strip line structure. In Fig. **1a**, the size of the prototype was 75mm×10mm×2mm.

Hirokazu Tohya

In Fig. **1b**, the prototype was formed by the strip line which consists of three ceramics having 50μm thickness and the metal layer which was formed by the silk screen. Two thick ceramics were used at the outside of the strip line for increasing its strength. The ceramics for the strip line consisted of the materials of Pb (Zr, Ti) 03 system.

(a) Outer Appearance	(b) Layer Formation

Figure 1: Prototype Chip Ceramic LILC of Strip Line Structure.

Fig. **2** show the measured dielectric constant and the dielectric loss tanδ of the insulator of the prototype.

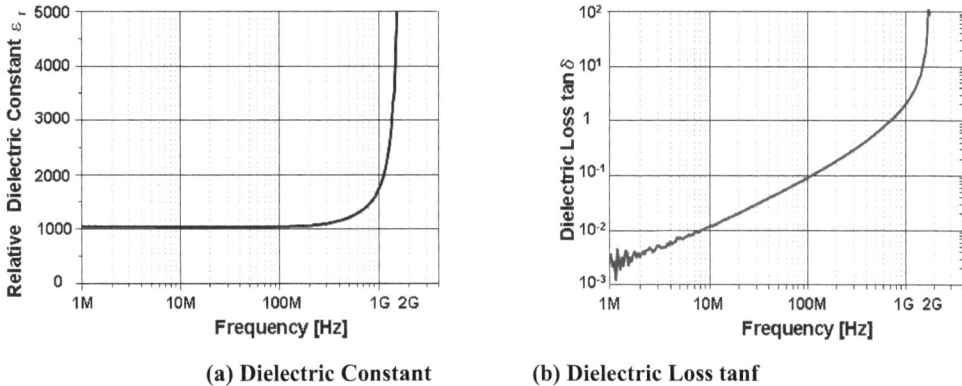

(a) Dielectric Constant	(b) Dielectric Loss tanf

Figure 2: Measured Characteristics of Insulator of Prototype.

Fig. **3** show the measured transmission coefficient S_{21} of the prototype of the chip ceramic LILC. The measured S_{21} is the satisfying value but the size is too large to use as the decoupling component on the common PCB.

Fig. **4** shows the prototype of the multi-layers chip ceramic LILC. In Fig. **4a**, the size of the prototype was 5.7mm×5mm×2mm. In Fig. **4b**, the prototype consisted of 25 layers of the ceramics and the metal plated ceramics, and it was formed switchback transmission line by the vias.

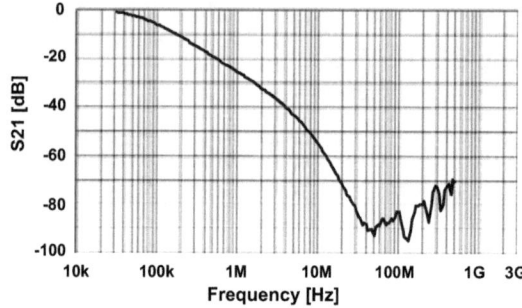

Figure 3: Measured S_{21} of Prototype Chip Ceramic LILC.

(a) Outer Appearance

(b) Layer Formation

Figure 4: Prototype Multi-layers Chip Ceramic LILC.

Fig. **5** shows the measured transmission coefficient S_{21} of the prototype of the multi-layer chip ceramic LILC. The measured S_{21} is the satisfying value but the formation is too complex and the manufacturing cost was estimated to be too expensive. The capacitance of the prototype of the multi-layer chip ceramic LILC was $220nF$ and S_{21} at $1GHz$ is about 20dB smaller than it of the chip ceramic capacitor of $100nF$.

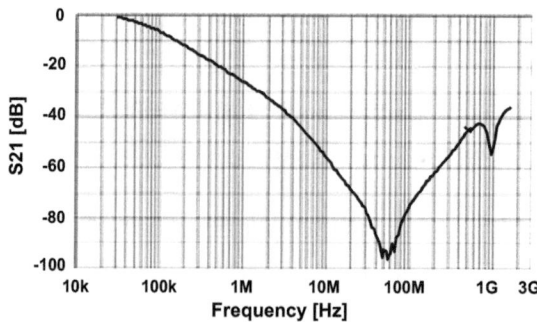

Figure 5: Measured S_{21} of Prototype Multi-layers Chip Ceramic LILC.

Fig. **6** shows the calculated conductivity of the adopted ceramics.

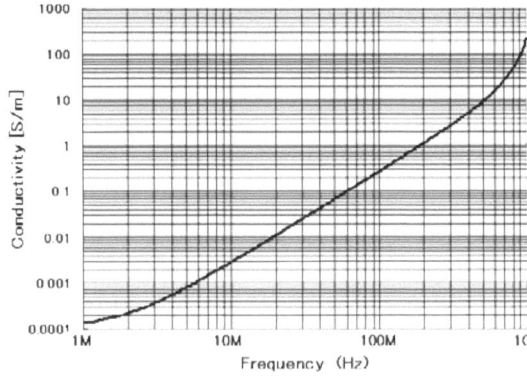

Figure 6: Calculated Conductivity of Adopted Ceramics.

In Fig. **6**, the conductivity was calculated by using the measured dielectric constant and the loss tangent in accordance with the definition of the loss transient. Unfortunately, the conductivity of the adopted ceramics was not suitable as the insulator of the transmission line.

CERAMIC CYLINDER LILC

The ceramic cylinder LILC was discussed because the ceramic cylinder capacitor had been used for the radio receiver as the high-frequency capacitor in the past.

Fig. **7** shows the prototype of the ceramic cylinder LILC. In Fig. **7b**, the ceramics was same as it which was adopted to the chip ceramic LILC, each L1 and L2 of the long type was 100*mm* and 88.6*mm*, and each L1 and L2 of the short type was 5.4*mm* and 4.6*mm*, D1 was 1.8*mm*, and the thickness of the ceramics was 0.14*mm*, the conductor was formed by the silver paste.

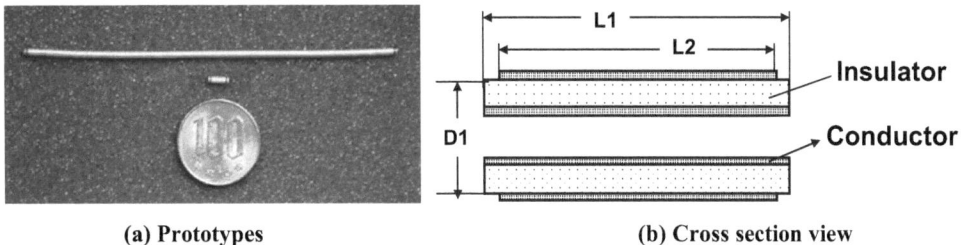

(a) Prototypes   (b) Cross section view

Figure 7: Prototypes of Ceramic Cylinder LILC.

Fig. **8** shows the measured S_{21} of the prototypes of the ceramic cylinder LILC. Each S_{21} of the long type and the short type of LILC at $1GHz$ is about 68dB and 25dB smaller than it of the chip ceramic capacitor. This measured value was considered to be suitable for the high-performance decoupling component.

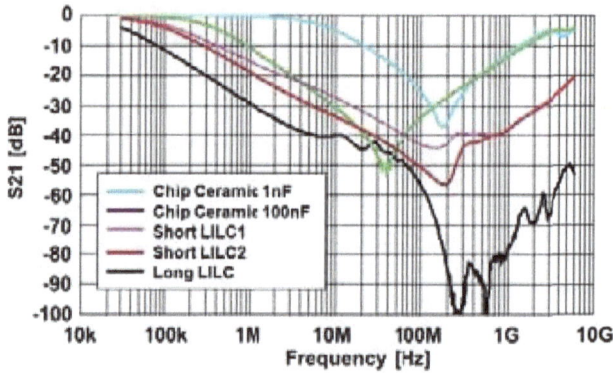

Figure 8: Measured S_{21} of Prototype of Ceramic Cylinder LILC.

Fig. **9** shows the test board for the short type of the ceramic cylinder LILC. The size of the board which consists of four layers was $300mm \times 300mm \times 1.6mm$. PCB consisted of a signal trace, a ground plane, a power plane, and a signal trace. The FPGA which includes $60MHz$ ring oscillators and nineteen pieces of the short ceramic cylinder LILC were used. Two edges of the inner conductor and the outer conductor of the ceramic cylinder LILC were connected to the trace of the PCB by the solder.

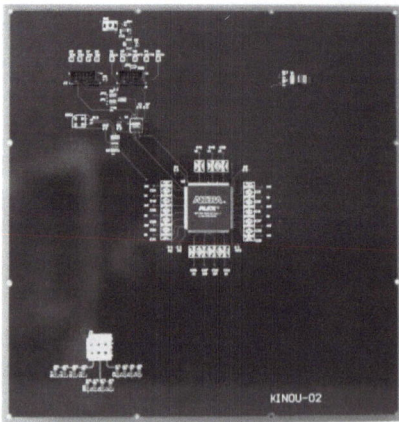

(a) Cylinder LILC on Test Board   (b) Close-up view

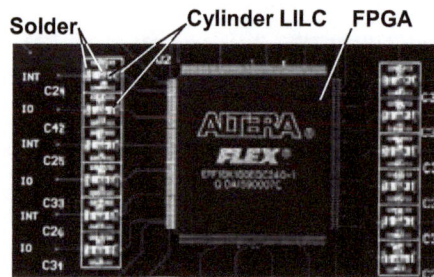

Figure 9: Test Board for Short Type of Ceramic Cylinder LILC.

Fig. **10** shows the test board for the long ceramic cylinder LILC. The size of the board, the layer formation, and the FPGA were same as the test board shown in Fig. **9**, and nineteen long cylinder LILC were used, and two edges of the inner conductor and the ten points of the outer conductor of the ceramic cylinder LILC were connected to the trace of the PCB by the solder because it was not the straight-line.

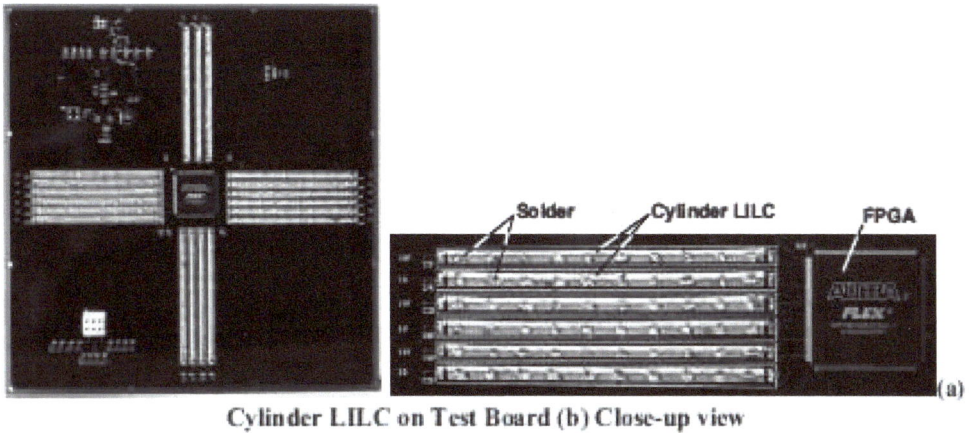

Cylinder LILC on Test Board (b) Close-up view

Figure 10: Test Board for Long Type of Ceramic Cylinder LILC.

Fig. 11 shows the measured decoupling performance of the ceramic cylinder LILC on the board.

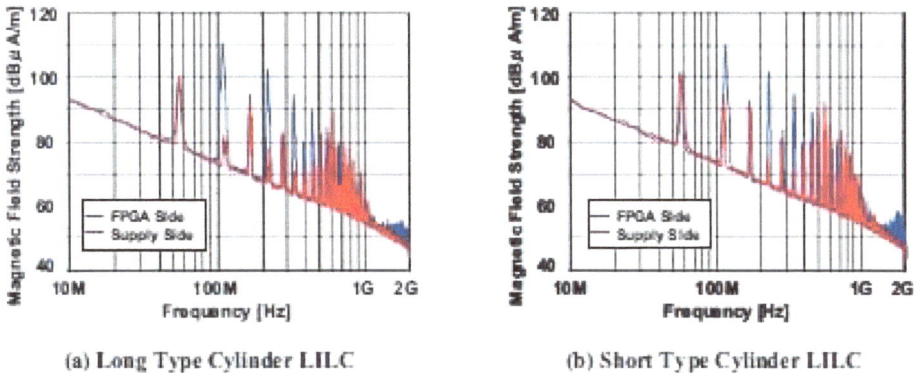

(a) Long Type Cylinder LILC   (b) Short Type Cylinder LILC

Figure 11: Decoupling Performance of Prototype of Ceramic Cylinder LILC on Test Board.

The attenuation of the long type is relatively large and the long type is too brittle to manufacture and the application. In addition, the attenuation performance of the long type is not suitable for decoupling on PCB. And, the conductivity of the adopted ceramics to the prototype of the cylinder LILC was not suitable as the insulator of it.

SOLID ALUMINUM LILC [1]

Fig. **12** shows the first prototype of the solid aluminum LILC.

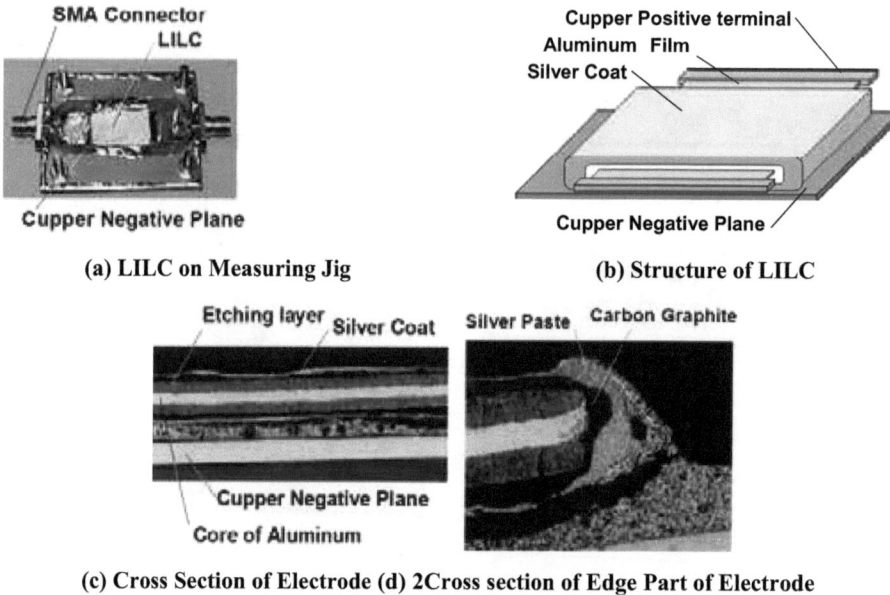

(a) LILC on Measuring Jig

(b) Structure of LILC

(c) Cross Section of Electrode (d) 2Cross section of Edge Part of Electrode

Figure 12: 1st Prototype of Solid Aluminum LILC.

In Fig. **12**, the chip of LILC consisted of the core layer of the $50\mu m$ thickness and the etched layer which is formed at both side of the aluminum film. The thickness and the dielectric constant of the insulator layer which is formed on the etched layer were $50nm$ and 8. Each thickness and conductivity of the conductive polymer layer which was formed on the insulator layer was $3\mu m$ and $3,000S/m$. Each thickness and conductivity of the carbon paste layer which is formed on the conductive polymer layer was $30\mu m$ and $30,000S/m$. The thickness of the silver coat layer which is formed on the carbon paste layer was $12\mu m$. There is not the etched layer on the edge face of the aluminum film but there are only the insulator layer and the carbon paste layer on it.

Fig. **13** shows the image of the chip formation of LILC. The chip of LILC was formed by the etched aluminum film. The conductive polymer layer was formed on the surface of the etched aluminum film.

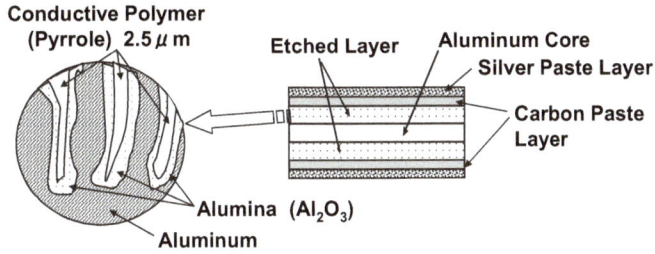

Figure 13: Image of Chip Formation of LILC.

The carbon paste layer was formed on the surface of the conductive polymer layer. The silver paste layer was formed to the surface of the carbon paste. The expanded area of the surface of the etched aluminum film was hundredfold approximately.

Fig. **14** shows the measured S_{21} of the first Prototypes of LILC.

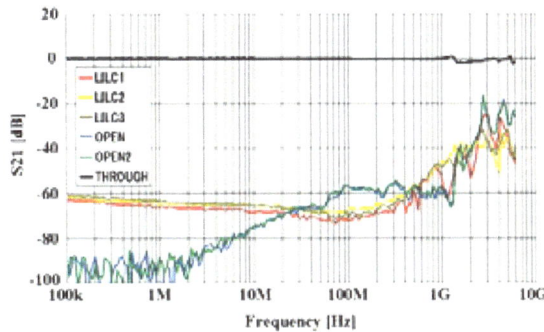

Figure 14: Measured S_{21} of 1st Prototypes of LILC.

In Fig. **14**, "OPEN" means the non-connection state of the connectors of the equipped cables of the network analyzer. "THROUGH" means the mutual connection state of the connectors of the equipped cables of the network analyzer. The terminal impedance was calculated by the transmission coefficient which is the conventional method used to the decoupling capacitor. However, afterward, it was understood that this method is incorrect in the case of the LILC.

From above study result, the component which has the formation of the transmission line was suitable for the decoupling component because the transmission line can reflect and attenuate EMW on PDN effectively. The transmission line consists of two metal layers and the insulator which is taken between two metal layers and this formation is similar to it of the capacitor. The best formation of the electrode was considered to be it of the solid aluminum capacitor from the points of view of the manufacturability, availability, and the cost. Therefor the improvement of the solid aluminum LILC was started.

REFERENCES

[1]   K. Masuda, H. Tohya, M. Satoh, "A Shielded Strip Type Low Impedance Line Component Using a Conducting Polymer for Wide Frequency Band De-Coupler Set", IEICE Transactions on Electronics, vol. E85-C, No. 6, pp. 1317-1322, 2002.

Send Orders for Reprints to reprints@benthamscience.net

Switching Mode Circuit Analysis and Design, 2013, 109-119   109

CHAPTER 8

Improvement of Solid Aluminum LILC [1, 2]

Abstract: The best decoupling component was the solid aluminum LILC. The prototyping result for solving the problems which was clarified at the feasibility study is presented.

Keywords: Solid aluminum LILC, slender chip formation, chip length, aluminum film, alumina, universal test fixture, S parameter, test jig, transmission coefficient, electromagnetic coupling, roll-roll manufacturing, chip length, attenuation constant, terminal impedance, FDTD simulator, transmission characteristics, poly pyrrole (Ppy), PEDOT, TCNQ, current capability.

PROTOTYPING

S_{21} of the solid aluminum LILC depends on its length at the high-frequency over $10MHz$ and it depends on the capacitance at the lower frequency than $10MHz$ approximately. The large capacitance can be got easily by the current cheap capacitor such as the solid or liquid aluminum capacitor. Therefore, the slender chip formation was considered to be reasonable as the improved solid aluminum LILC.

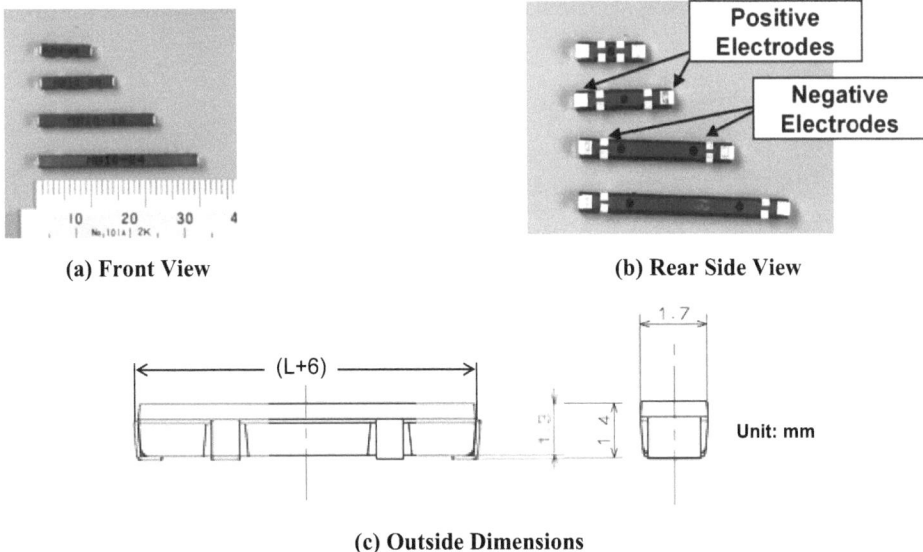

(a) Front View   (b) Rear Side View

(c) Outside Dimensions

Figure 1: Final Stage Prototype of Solid Aluminum LILC.

Hirokazu Tohya

Fig. **1** shows the final prototypes of the solid aluminum LILC. Each thickness and width of the chip was 0.23*mm* and 1.5*mm*. L shown in Fig. **23c** means the chip length of the solid aluminum LILC. 1.7*mm* width responded to the chip width of 1*mm*. The thickness of 0.23mm was considered to be too thick for the roll-roll manufacturing.

Fig. **2** shows the example of the bonding feature of the terminal of the chip. The difficulty of the process condition of the bonding is relatively high, because the surface of the aluminum film is coated by the alumina rapidly at the high-temperature environment.

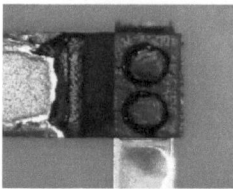

(a) Laser Irradiation Side   (b) Reverse Side

Figure 2: Example of Bonding Feature of Terminal of Chip.

Fig. **3** shows the universal test fixture for measuring the *S* parameter of the chip.

Figure 3: Universal Test Fixture for Measuring S-Parameter of Chip.

Fig. **4** shows the measured S_{21} of the best samples of the prototypes of the LILC chip by using the universal test fixture shown in Fig. **3**. The chip width was 1.5*mm*, and S_{21} is greatly smaller than it of the capacitors and the other type of LILC.

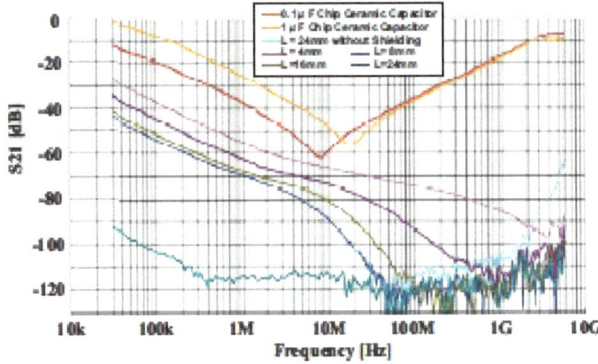

Figure 4: Measured S_{21} of Best Samples of Prototypes of LILC Chip.

Fig. 5 shows an example of the test jig for measurement of the transmission coefficient (S_{21}) of the LILC.

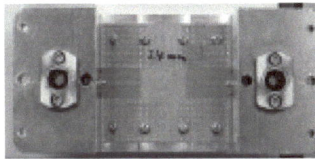

Figure 5: Example of Test Jig for Measurement of Transmission Coefficient of LILC.

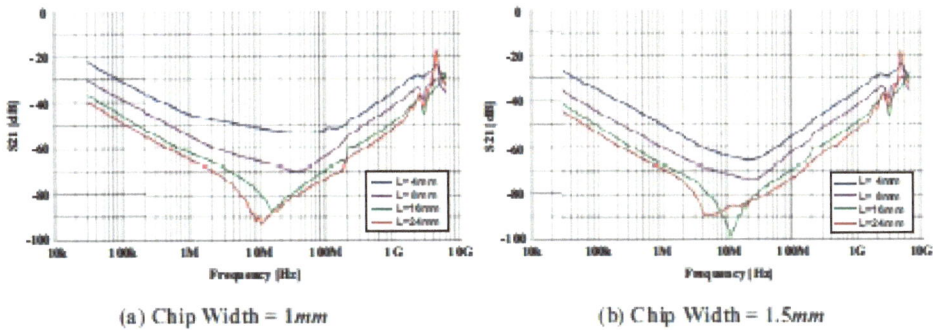

(a) Chip Width = 1*mm*   (b) Chip Width = 1.5*mm*

Figure 6: Measured S_{21} of Best Samples of Prototypes of Solid Aluminum LILC.

Fig. **6** shows the measured S_{21} of the best samples of the prototypes of the solid aluminum LILC by using the test jig. Fig. **6a** shows S_{21} of the 1*mm* chip width and Fig. **6b** shows S_{21} of the 1.5*mm* chip width. S_{21} of the LILC is larger than it of the chip of the LILC at the frequency higher than 100*MHz* approximately because the influence of the electromagnetic coupling between the positive electrodes increases. S_{21} in Fig. **6a** and **6b** is similar at the frequency higher than $100MHz$ approximately in spite of the different chip width. S_{21} at the frequency higher than $100MHz$ approximately is depend on the chip length.

ANALYSIS OF TERMINAL IMPEDANCE

Fig. **7** shows the cross section of the LILC chip. The conductive polymer was filled up in the etched layers. The surface of the LILC chip was coated by the silver paste A. One side of the LILC chip was attached to the copper plate having the $100\mu m$ thickness by the silver paste B. The copper plate was formed to the negative electrode and the two edges of the aluminum core were formed to the positive electrode.

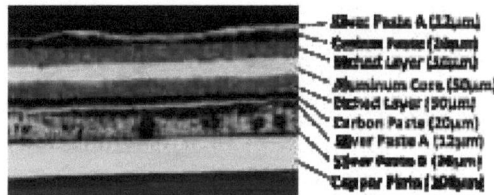

Figure 7: Cross Section of LILC chip.

Fig. 8 shows the model and the parameter of the material.

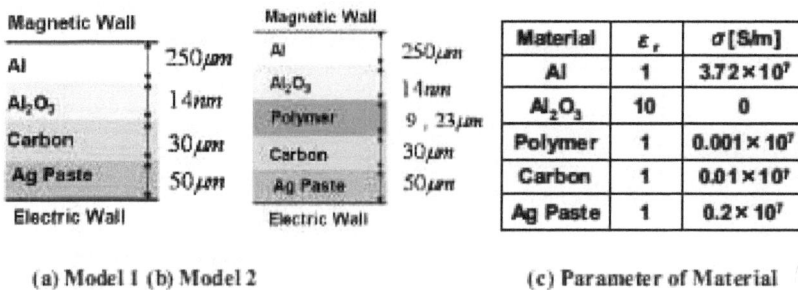

(a) Model 1 (b) Model 2   (c) Parameter of Material

Figure 8: Model and Parameter of Material.

In Fig. **8a**, the conductive polymer layer is not being formed. In Fig. **8b**, the etched layer was replaced to the conductive polymer layer because the formation of the etched layer of the aluminum film is too complex to form the model. In Fig. **8c**, the conductivity of the carbon graphite was adopted to the carbon paste layer.

The current of the transmission line is:

$$\dot{I} = i(t) = \oint H dl = H_x(y = h_2) - H_x(y = h_1) \tag{1}$$

The power of the transmission line is:

$$\dot{P} = p(t) = \int_{y=0}^{y=h_3} E_y H_x \cdot dy \tag{2}$$

where h_1 is the height from the surface of the silver paste layer to the surface of the aluminum, h_2 is the height from the surface of the silver paste layer to another surface of the aluminum, and h_3 is the height from the surface of the silver paste layer to another surface of the silver paste.

The impedance is:

$$\dot{Z} = \dot{P}/\dot{I}^2 \tag{3}$$

Fig. **9** shows the calculated terminal impedance by the equation 3. The line width was 1.5mm. Each "$9\mu m$" and "$23\mu m$" of the polymer layer in the model 2 shows the thickness. The FDTD simulator was used to the calculation. The calculated impedance was small enough at the broad band.

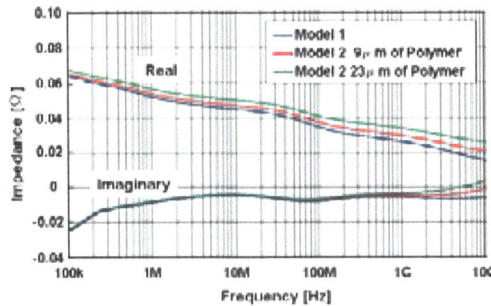

Figure 9: Calculated Terminal Impedance.

ANALYSIS OF TRANSMISSION COEFFICIENT (S_{21})

Fig. **10** shows the calculated attenuation constant α. The attenuation constant was calculated by the FDTD simulator by using the model shown in Fig. **8**.

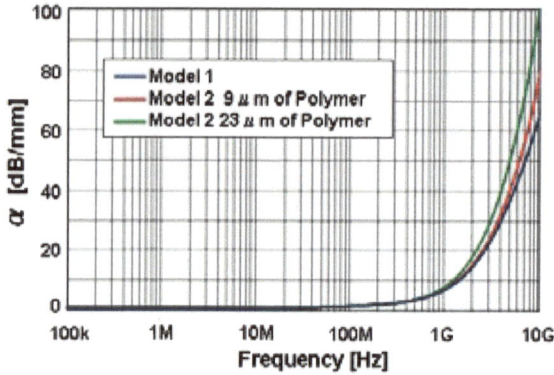

Figure 10: Calculated Attenuation Constant.

Fig. **11** shows the calculated transmission characteristics consists of the reflection coefficient S_{11} and the transmission coefficient S_{21} of the chip of LILC. The line length x was 16*mm* and the line width was 1.5*mm*. The calculated S_{21} is larger than the measured S_{21} because the etching effect was neglected.

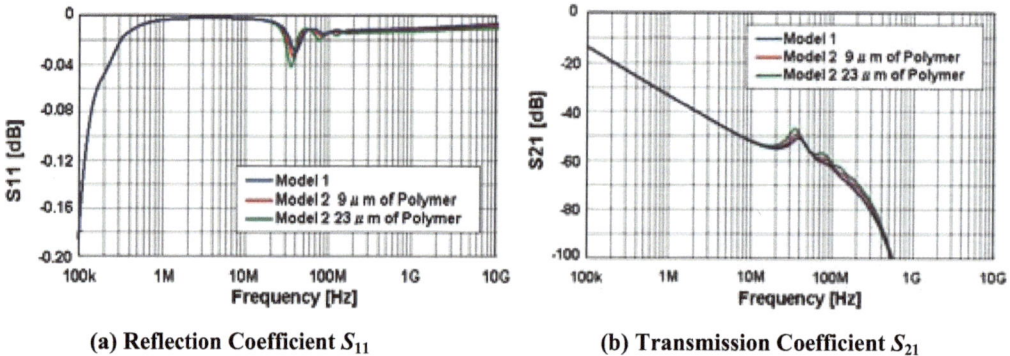

    (a) Reflection Coefficient S_{11}   **(b) Transmission Coefficient S_{21}**

Figure 11: Calculated Transmission Characteristics of 16*mm* Chip of LILC.

OPTIMIZATION OF CONDUCTIVE POLYMER AND ETCHING LAYER

Fig. **12** shows the measured S_{21} of the typical samples of the second group of the 24*mm* prototypes. S_{21} is little influenced by the thickness of the conductive polymer which consisted of the poly pyrrole (Ppy). S_{21} depends on the capacitance of the LILC.

Fig. **13** shows the measured S_{21} of the typical samples of the 3rd group of the 24*mm* prototypes of the LILC. All thickness of all samples was 23μm, the PEDOT is a conductive polymer based on 3,4-ethylenedioxythiophene, and TCNQ is the abbreviation of tetracyanoquinodimethane and it is also the conductive polymer. Ppy of $3000S/m$ was got the best characteristics of S_{21} characteristics at over than $1MHz$, PEDOT was got the best characteristics of S_{21} characteristics at up to $1MHz$.

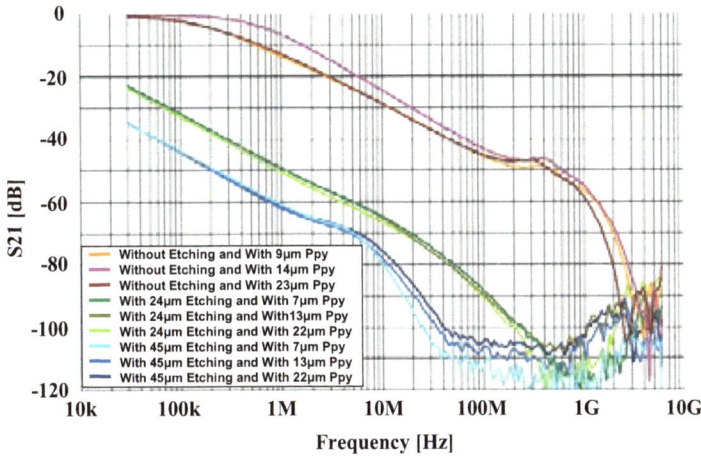

Figure 12: Measured S_{21} of Typical Samples of 2nd Group of $24mm$ Prototypes of LILC.

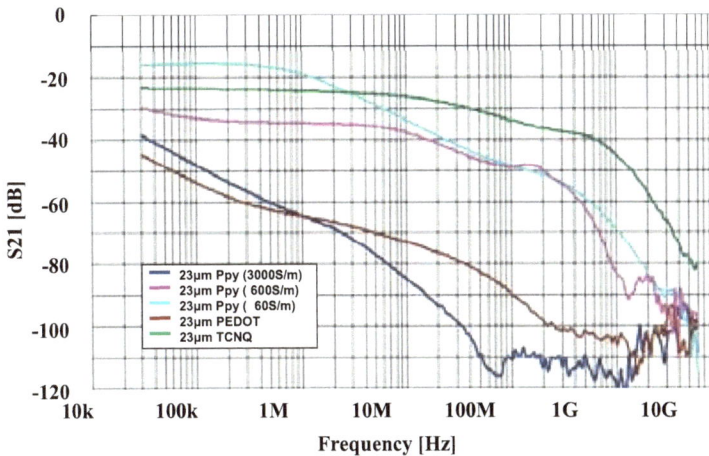

Figure 13: Measured S_{21} of Typical Samples of 3rd Group of $24mm$ Prototypes of LILC.

Fig. **14** shows the measured S_{11} of the other typical samples of the 2nd group of the prototypes of LILC having the $24mm$ chip length.

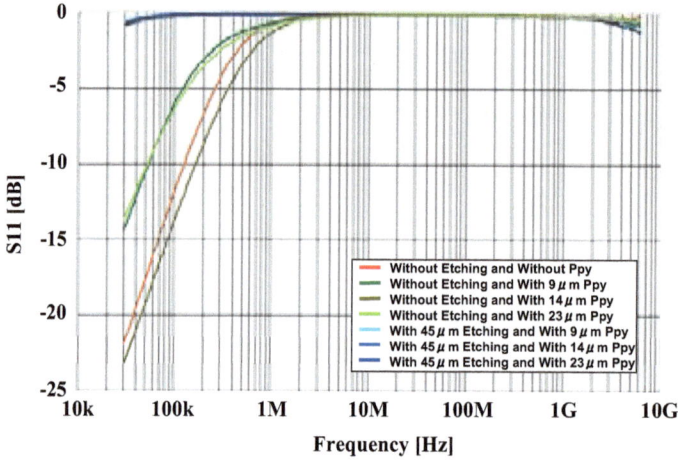

Figure 14: Measured S_{11} of Typical Samples of 2nd Group of Prototypes of 24*mm* LILC.

The relationship between the terminal impedance Z_1 and the measured S_{11} is:

$$Z_1 = \frac{1-S_{11}}{1+S_{11}} Z_0 \qquad (4)$$

where Z_0 is 50Ω.

Fig. **15** shows the calculated terminal impedance of the typical samples of the 2nd group of the prototypes of LILC having the 24*mm* chip length. The terminal impedance was calculated from S_{11} in Fig. **14** by the equation 4 and the limit of the accuracy depends on the network analyzer.

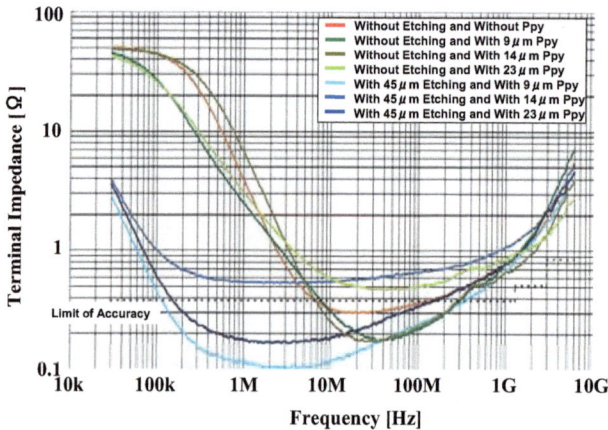

Figure 15: Calculated Terminal Impedance of 2nd Group of Prototypes of 24*mm* LILC.

From above discussion, the thick etching layer and the thick Ppy layer were considered to be suitable for the LILC. In addition, it was confirmed to be difficult to get the terminal impedance of the improved LILC from the measured S_{11} due to the limit of the accuracy of the current network analyzer.

ELECTRIC CURRENT CAPABILITY

Fig. **16** shows the typical application of LILC on the PCB.

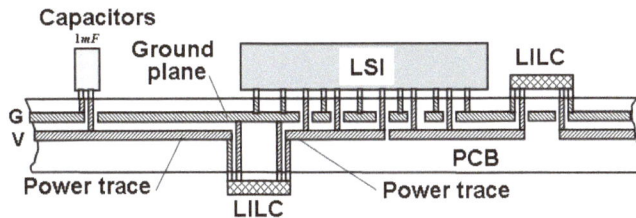

Figure 16: Typical Configuration of the application of LILC.

In Fig. **16**, LILC is bridging the power trace at under the LSI of the PCB. Two ground terminals of LILC are connected to the ground plane of the PCB. Therefore, the electric current does not flow in the conductor of the negative electrode. However, the electric current flows in the aluminum core of the positive electrode. The conventional capacitors are being used with LILC. These capacitors have large capacitance and they protect LSI from the instantaneous interruption of the power supply.

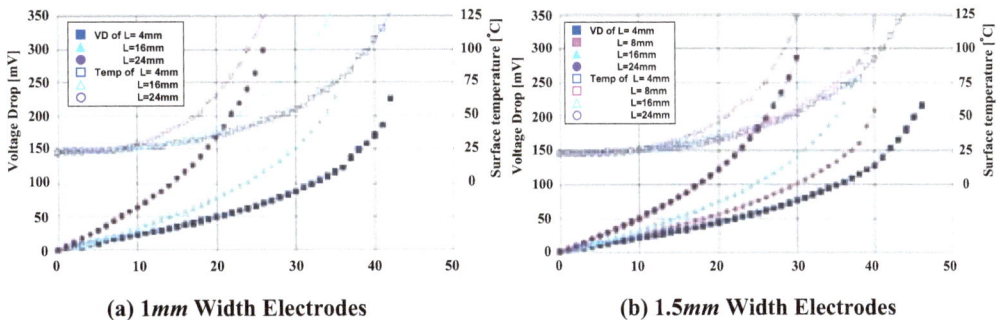

(a) **1*mm* Width Electrodes**   (b) **1.5*mm* Width Electrodes**

Figure 17: Measured Voltage Drop and Surface Temperature of Improved LILC.

As above mentioned, unlike the conventional capacitor, the LILC has the rated DC current. Fig. **17** shows the measured voltage drop and the surface temperature of the improved LILC. Fig. **17a** shows the voltage drop of the 1*mm* chip width and Fig. **17b** shows the voltage drop of the1.5*mm* chip width. In Fig. **17a**, the electrode was formed by the phosphor bronze film of the 1mm width and the 0.23*mm* thickness. In Fig. **17b**, the electrode was formed by the phosphor bronze film of the 1.5*mm* width and the 0.23*mm* thickness.

From the test result, when the voltage drop is selected to 50*mV* and the chip width is 1mm, the maximum electric current is 7*A* at the 4*mm* chip length, it is 14*A* at the 16*mm* chip length, and it is 20*A* at the 24*mm* chip length. When the voltage drop is selected to 50*mV* and the chip width is 1.5*mm*, the maximum electric current is 10*A* at the 4*mm* chip length, it is 17*A* at the 16*mm* chip length, and it is 22*A* at the 24*mm* chip length.

CONCLUSIONS

From above discussion in this chapter, the solid aluminum LILC was confirmed to be the most suitable decoupling component. However some important problems existed in order to commercialize.

The confirmed major problems are as follows;

a. To develop the analyzing method for improving the accuracy of the calculation of the transmission characteristics.

b. Reduction of the thickness of the aluminum film of the chip for the efficient manufacturing.

c. The improvement of the reliability of the bonding between the aluminum film and the lead-frames.

d. The capable DC current is limited by the cross-section squire of the aluminum film.

e. To eliminate the need for the chemical reaction at the chip manufacturing.

f. To clarify the mechanism of the influence to the operation stability of the transistor or LSI.

g.   To develop the application technology on PCB and on the chip of LSI.

REFERENCES

[1]   Hirokazu TOHYA, Koichiro MASUDA, Hideki SHIMIZU, "Low Impedance Line structure Component (LILC) Technology well exceeding the performance of Capacitor" IEICE, TECHNICAL REPORT OF IEICE, EE 2002-29, 2002.

[2]   Hirokazu TOHYA, Koichiro MASUDA, Hideki Shimizu and Yoshiaki WAKABAYASHI, "Low Impedance Lune Structure Component (LILC) for Power Distribution System in the High-Performance Digital Circuit", 2003 Southwest Symposium on Mixed-Signal Design in Las Vegas, vol. 1, pp. 60-65, February 23-25, 2003.

Send Orders for Reprints to reprints@benthamscience.net

CHAPTER 9

Novel Characteristic Equations for Decoupling Component

Abstract: For the effective improvement of the solid aluminum LILC, the characteristic equations which have the higher accuracy and convenience than the conventional EDTD simulators were developed.

Keywords: Characteristic equations, decoupling component, PDN, FDTD, broadband pulse, Gaussian pulse, grid spatial discretization, computational domains, Maxwell's equations, EMW theory, aluminum film, alumina film, etching formation, appearance ratio, effective thickness, lossy material layer, skin depth, electromagnetic coupling, transmission coefficient, terminal impedance.

PROBLEM OF ANALYZING METHOD OF LILC

The high-performance decoupling component which works on PDN is one of the requisite components for forming QSCC of SMC. LILC was analyzed by the conventional FDTD simulation method. Every modeling technique has strengths and weaknesses, and the FDTD method is no different. According to Wikipedia, the strength of the FDTD method for analyzing LILC is the following;

1. FDTD is a versatile modeling technique used to solve Maxwell's equations. It is intuitive, so users can easily understand how to use it and know what to expect from a given model.

2. FDTD is a time-domain technique. When a broadband pulse (such as a Gaussian pulse) is used as the source, the response of the system over a wide range of frequencies can be obtained with a single simulation. This is useful in applications where resonant frequencies are not exactly known, or anytime that a broadband result is desired.

3. The FDTD technique allows the user to specify the material at all points within the computational domain. A wide variety of linear and nonlinear dielectric and magnetic materials can be naturally and easily modeled.

4. FDTD allows the effects of apertures to be determined directly. Shielding effects can be found, and the fields both inside and outside a structure can be found directly or indirectly.

Hirokazu Tohya

In contrast, according to Wikipedia, the weakness of the FDTD method for analyzing LILC is the following;

1. Since FDTD requires that the entire computational domain be gridded, and the grid spatial discretization must be sufficiently fine to resolve both the smallest wave length and the smallest geometrical feature in the model, very large computational domains can be developed, which results in very long solution times. The models with long and thin features are difficult to the model in FDTD because of the excessively large computational domain required.

2. There is no way to determine unique values for permittivity and permeability at a material interface.

LILC consists of the etched aluminum film coated by the alumina film. The etching formation is quite complex. The surface of the etching formation is expanded to more than hundredfold of the plane surface. The diameter of the etching pit of the surface is several micrometers and it of the bottom is a few nanometers. In addition, the thickness of the alumina film is a few ten nanometers. The differences between these sizes and the external size of LILC are great. From above, the simulation of the characteristics of LILC for comparison with the measured data is considered to be impossible. The solution of the mentioned problems of LILC is impossible if this situation goes on.

Therefore, the novel characteristic equation which does not use the grid of the computational domain became necessary. They are based on the Maxwell's equations and his EMW theory.

CHARACTERISTIC EQUATIONS [1, 2]

The extension ratio of the etched aluminum layer is:

$$k = \frac{C_1 \cdot a}{\varepsilon_0 \varepsilon_r \cdot 2 \cdot 10^{-4}} \tag{1}$$

where C_1 is the capacitance per $1 cm^2$ of the etched aluminum film, a is the thickness of the alumina on the etching surface.

The impedance of the capacitor of the chip of the decoupling component is:

$$Z_c = \frac{1}{z \cdot w \cdot 2\pi f C_1 \cdot 10^4} \tag{2}$$

where z is the chip length, and w is the chip width.

The approximate characteristics of the terminal impedance of the transmission line having the finite length is:

$$Z_a = Z_C + Z_0 \tag{3}$$

where Z_0 is the characteristic impedance of the transmission line of the chip of the decoupling component.

The chip of the decoupling component is formed by two parallel plates having the etched face. The characteristic impedance is:

$$Z_{L0} = \frac{h_e}{\sqrt{R} \cdot 2w} \sqrt{\frac{\mu_0}{\varepsilon_r \varepsilon_0}} \tag{4}$$

where R is the appearance ratio of the capacitance, and h_e is the effective thickness of the insulator.

The effective thickness of the insulator depends on the intrinsic impedance and the thickness of the materials which are the alumina, conductive polymer, carbon paste, and the air-gap between the aluminum and the silver coating.

The carbon paste layer consists of the carbon graphite particles and the binder. The effective thickness of the carbon paste layer is:

$$b_{ec} = b_c \frac{1}{\sigma_c \cdot \rho_c} \tag{5}$$

where b_C is the thickness of the carbon paste layer, ρ_C is the resistivity of the carbon paste, and σ_C is the conductivity of the carbon graphite.

The air-gap layer consists of the microscopic cavity in the etching layer. Therefore, the equivalent thickness b_A is used for the air-gap layer. The functional sinking of the electric field and magnetic field when EMW travels on the transmission line can be considered to depend on the intrinsic impedance of each material against it of the alumina which is the insulator. The effective thickness of the insulator layers between the silver coating layer and the aluminum layer is:

$$h_e = \frac{a Z_I + b_S Z_S + b_{eC} Z_{Ca} + b_A Z_A}{Z_I} \tag{6}$$

where b_S is the thickness of the conductive polymer layer. Each Z_I, Z_S, Z_{Ca}, and Z_A is the intrinsic impedance of the alumina, it of the conductive polymer, it of the carbon graphite, and it of the air.

The electric field and magnetic field which sink to the lossy material when EMW travels on the transmission line is:

$$E_x = E_{x0} e^{-z/\delta}, \; H_y = H_{y0} e^{-z/\delta} = \sigma \delta E_{x0} e^{-z/\delta} \tag{7}$$

The electric power which is consumed in the minute volume is:

$$p \, dxdydz = E_x dx \, H_y dy \, dz = \sigma \delta E_{x0}{}^2 e^{-2z/\delta} \, dz \tag{8}$$

When the thickness of the conductor is infinite, the electric power which is consumed per unit square of the surface of the conductor is:

$$P_L = \int_0^\infty p \, dz = \int_0^\infty \sigma \delta |E_{x0}|^2 \, e^{-2z/\delta} dz = \sigma \delta |E_{x0}|^2 = \frac{1}{R} |E_{x0}|^2 \tag{9}$$

where z is the thickness of the lossy material layer, δ is the skin depth of the conductor.

From the attenuation constant of the characteristics of the transmission line and the equation 9, the attenuation constant by the lossy material layer which forms the transmission line is:

$$\alpha = 2 \frac{R\sqrt{C/L}}{2} = \frac{1}{\delta \sigma w Z_0} \tag{10}$$

The lossy material layer of the chip of the decoupling component exists at one face of the transmission line. From the equation 10, the effective attenuation constant at each corresponding layer of the chip of the decoupling component is:

$$\alpha_n = \frac{A_n}{2 \cdot Z_0 wR\sqrt{k} \delta_n \sigma_n} \tag{11}$$

where each $\delta_n = 1/\sqrt{\pi f \mu_0 \sigma_n}$ and σ_n is the skin depth and the conductivity of the conductive polymer and the carbon graphite.

The transmission loss appears at the conductive polymer layer and the carbon graphite layer when EMW travels in the chip of the decoupling component. In the equation 11, the proportion of the loss at the finite depth to infinite depth of each layer is:

$$A_n = 1 - e^{-b_n/\delta_n} \tag{12}$$

where b_n is the thickness of the conductive polymer and the effective thickness of the carbon paste layer.

Transmission coefficient of the chip of the decoupling component which consists of two parallel plates is:

$$S_{21\alpha} = e^{-\sum_1^n \alpha_n \cdot z \cdot R\sqrt{k} \cdot F} \tag{13}$$

where F is the fractional shortening.

When a couple of the terminals are connected to the infinite transmission lines, the transmission coefficient of the chip of the decoupling component is:

$$S_{21RM} = 1 - \left(\frac{Z_{L0} - Z_0}{Z_{L0} + Z_0}\right)^2 \tag{14}$$

The electromagnetic coupling is considered to be being between the terminals of the chip of the decoupling component. This can be handled as the capacitance coupling or the inductance coupling when the distance between terminals is shorter than the wavelength.

The impedance of the coupling capacitance between terminals of the decoupling component is:

$$Z_{CT} = \frac{1}{2\pi f C_T} \tag{15}$$

where C_T is the coupling capacitance.

The transmission coefficient of the electromagnetic coupling between terminals of the decoupling component is:

$$S_{21T} = \frac{2}{(Z_{CT}/Z_0)+2} \tag{16}$$

The full-wave transmission coefficient of the chip of the decoupling component is:

$$S_{21R} = S_{21R} + S_{21RM} \tag{17}$$

The transmission coefficient of chip of the decoupling component is:

$$S_{21L} = S_{21\alpha} + S_{21R} \tag{18}$$

The approximate transmission coefficient adjusted by S_{21T} of the decoupling component is:

$$S_{21LM} = \sqrt{S_{21L}^2 + S_{21T}^2} \tag{19}$$

The terminal impedance of the decoupling component is:

$$Z_L = \sqrt{Z_{L0}^2 + Z_C^2} \tag{20}$$

The terminal impedance of the decoupling component adjusted by Z_{CT} of the decoupling component is:

$$Z_{LM} = Z_L \left(1 + \frac{120\pi}{1+Z_{CT}}\right) \tag{21}$$

The impedance of the power distribution line of PCB is:

$$Z_{PD} = Z_{CP} + \frac{d_p\sqrt{\mu_0/\varepsilon_r\varepsilon_0}}{W_p} \tag{22}$$

where Z_{CP} is the summation of the capacitance of the power distribution line and the additional capacitor on the power distribution line, d_p is the distance between the power trace and the ground plane, and W_p is the width of the power trace.

The transmission coefficient of the decoupling component instead of the equation 30 when the decoupling component is mounted on the actual PCB is:

$$S_{21PD} = 1 - \left(\frac{Z_L - Z_{pD}}{Z_L + Z_{pD}}\right)^2 \tag{23}$$

The approximate transmission coefficient adjusted by S_{21T} of the decoupling component is:

$$S_{21PDM} = \sqrt{S_{21PD}^2 + S_{21T}^2} \tag{24}$$

The AC circuit theory cannot handle the traveling EMW because the circuit diagram in accordance with the AC circuit theory has no physical dimensions. When the capacitor is connected to the infinite transmission line in parallel, the transmission coefficient of this decoupling circuit is:

$$S_{21D} = \frac{2Z_{Ca}}{2Z_{Ca} + Z_0} \tag{25}$$

where Z_{Ca} is the capacitance of the decoupling capacitor.

The summation transmission coefficient of the transmission line connected the capacitor in parallel including the equivalent electromagnetic coupling is:

$$S_{21DM} = \sqrt{S_{21D}^2 + S_{21T}^2} \tag{26}$$

EVALUATION OF AVAILABILITY

The transmission coefficient S_{21} was calculated by the novel characteristic equation according to the parameters of the material and the process of the prototyped LILC. The formation of the chip of LILC was considered to be ideal because it is the shielded strip line similar to the coaxial. However, the calculation result was different largely from the measured value. It was doubted as this cause that the etching faces do not exist at the cut surface of the etched aluminum film.

Fig. **1** shows the calculated transmission coefficient and the terminal impedance of the prototype of LILC by using the recast calculation parameters and the novel characteristic equations. The calculation condition was the following; z of each LILC04, LILC08, LILC16, and LILC24 is 4mm, 8mm, 16mm, and 24mm, C_1 is 66μF, R is 1.0, a is 20.3nm, ρ_C is 33$\mu\Omega m$, each σ of the conductive polymer and the carbon graphite is 3000S/m and 72727S/m, μ_r of the alumina is 8.5, b_s is 3μm, b_c of each LILC04, LILC08, LILC16, and LILC24 is 0, 1μm, 40μm, and 50μm, b_A of LILC04 is 33nm, b$_a$ of LILC08, LILC16, and LILC24 is 0, w of LILC04 is 1.1mm, w of LILC08, LILC16, and LILC24 is 1.2mm, C_T of each LILC04, LILC08, LILC16, and LILC24 is $3.0\times10^{-14}F$, $1.333\times10^{-14}F$, $5.71\times10^{-15}F$, and $3.64\times10^{-15}F$, the thickness of the etched aluminum film is 0.25mm, and each k, R and F was 1.0, respectively. In calculation of the transmission coefficient and the characteristic impedance, the thickness of the etched aluminum film was used as the line width.

The calculated value shown in Fig. **1a** is well matched to the measured value in Fig. **6a** of Chapter **6**. In Fig. **1**, the transmission coefficient and terminal impedance of LILC04 is relatively larger than other LILC. The cause is considered to be the air gap b_a which exists only in LILC04. In Fig. **1b**, the measured value of the terminal impedance does not exist because too small to measure by the conventional network analyzer, however the calculation result will be reliable because the terminal impedance is used for the calculation of the transmission coefficient and its result was well matched to the measured value.

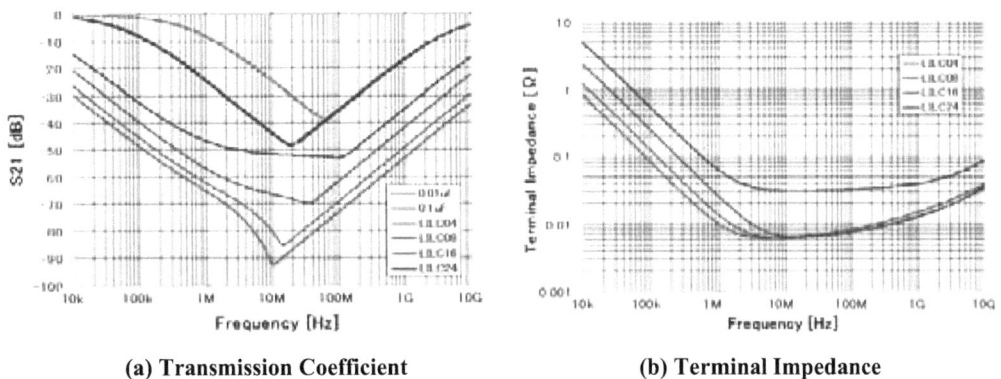

(a) Transmission Coefficient   (b) Terminal Impedance

Figure 1: Calculated Characteristics of Prototypes of Solid Aluminum LILC.

Fig. **2** shows the calculated transmission coefficient and the terminal impedance of the commercialized similar components.

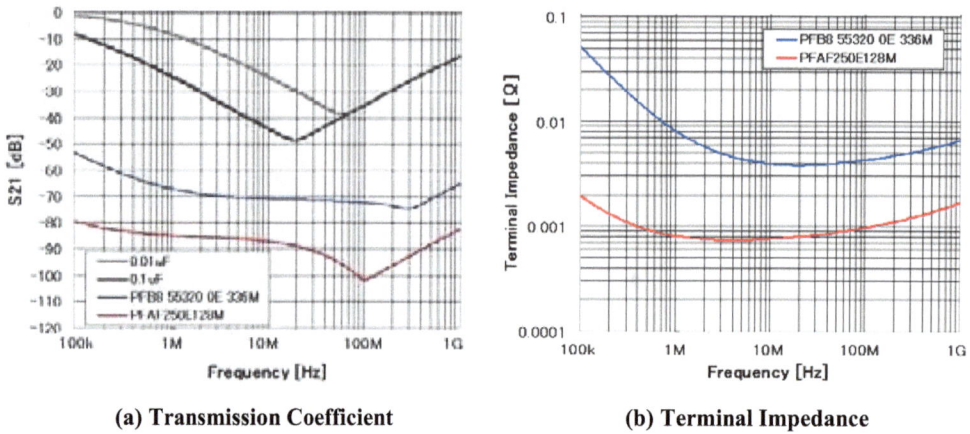

(a) Transmission Coefficient   (b) Terminal Impedance

Figure 2: Calculated Characteristics of Commercialized Similar Components.

In Fig. **2**, the calculation condition was the following; z of each PFB8 55320 0E 336M (336M) and PFAF250E128M (128M) was $4.6mm$ and $10.2mm$, C_1 of each 336M and 128M was $116\mu F$ and $216\mu F$, R was 1.0, a of each 336M and 128M was $9.70nm$ and $6.93nm$, ρ_C was $200\mu\Omega$, each σ of the conductive polymer and the carbon graphite was $3000S/m$ and $72727S/m$, μ_r of the alumina was 8.5, b_s was $1.2\mu m$, b_c was $5\mu m$, b_A was 0, w of each 336M and 128M was $1.1mm$ and $1.2\mu m$, C_T of each 336M and 128M was $0.2\times10^{-17}F$ and $0.1\times10^{-17}F$, the thickness of the etched aluminum film of each 336M and 128M was $0.1mm$ and $0.14mm$, and 336M consists of parallel connected two transmission lines and 128M consists of parallel connected 5 sub-assemblies of the chip, respectively. In calculation, the summation of the thickness of the etched aluminum film was used for the thickness of the etched aluminum film was used for w, and each k, R and F was 1.0, respectively.

When N times of transmission line are connected in parallel, the terminal impedance and attenuation constant become $1/N$ of the case of single transmission line. On the other hand, when N times of sub-assembly of the chip which are not coated by the silver paste but are coated by the carbon paste are connected in parallel, the width and the thickness of the chip become N times of the case of single transmission line. In calculation, the thickness of the etched aluminum film was used as the line width, and its extension ratio was 1.0. The calculated transmission coefficient of 336M and 128M is well matched to the measured it shown on the web site of TOKIN, and the reflection coefficient is considered to be dominant because the characteristic curve of the transmission coefficient is frat at the broad band. The calculated terminal impedance of 336M and 128M is small enough.

ESTIMATION OF CHARACTERISTICS OF LILC ON ACTUAL PCB

Fig. **3** shows the calculated transmission coefficient and the calculated terminal impedance when the prototypes of the solid aluminum LILC are mounted on the actual PCB.

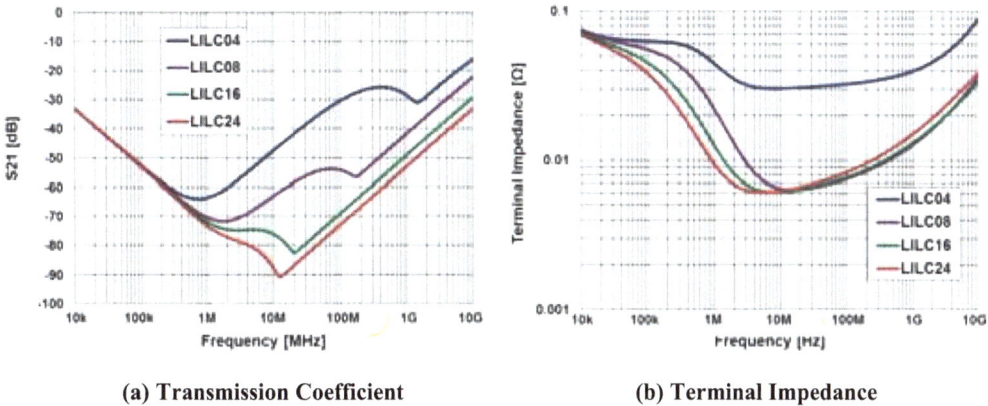

| (a) Transmission Coefficient | (b) Terminal Impedance |

Figure 3: Calculated Characteristics of Prototypes of Solid Aluminum LILC Used on PCB.

In Fig. **3**, the calculation condition is the following; the size of the power trace between the decoupling component and LSI was $40mm \times 40mm$ and the size of the power trace between the power source and the decoupling component was $10mm \times 100mm$, and the conventional capacitor which has capacitance of $1mF$, ESL is $15nH$, and ESR is $60m\Omega$ was connected in parallel. The calculated transmission coefficient is improved at the frequency band lower than $1MHz$ approximately, and the transmission coefficient is degraded at the frequency band higher than $10MHz$ approximately. The calculated terminal impedance is improved at the frequency band lower than $1MHz$ approximately.

Fig. **4** shows the calculated transmission coefficient and the calculated terminal impedance when the commercialized similar decoupling components are mounted on the actual PCB. The calculation condition was the following; the size of the power trace between the decoupling component and LSI was $100mm \times 100mm$ and the size of the power trace between the power source and the decoupling component was $100mm \times 200mm$. The transmission coefficient is being quite degraded at the all frequency band because the reflection constant is considered to be dominant. The terminal impedance is improved at the frequency band lower than $10MHz$ approximately because the capacitor of $1mF$ was connected in parallel, and the terminal impedance is not degraded at the frequency band higher than $10MHz$ approximately.

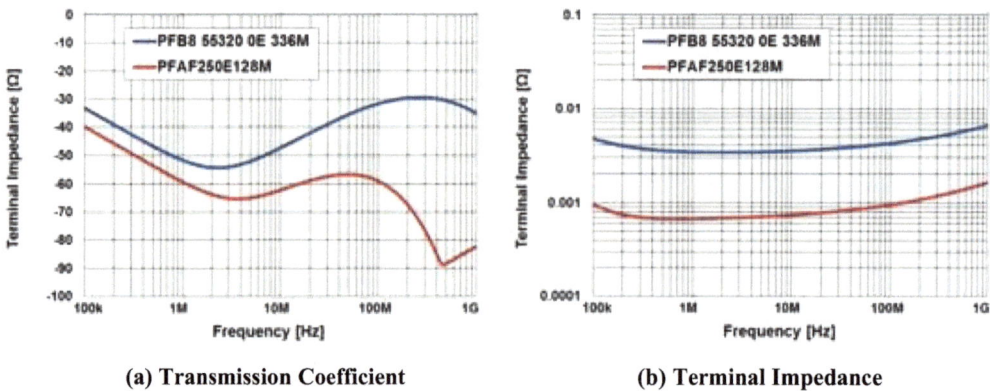

(a) Transmission Coefficient	(b) Terminal Impedance

Figure 4: Calculated Characteristics of Commercialized Similar Components Used on PCB.

From above, it was clarified that the novel characteristic equations based on the electromagnetism are more useful than the conventional simulators. The novel characteristic equations have been improved during development of the after-mentioned novel decoupling component. From the result of the review, it was clarified that the transmission coefficient of the prototyped LILC and the commercialized similar component depends on the transmission line formed by two cutting edges. Therefore the etched face has no effect for the characteristic impedance and the attenuation constant but has only effect for the capacitance. The variability of the thickness of the carbon paste layer is considered to be being influenced by the smooth surface of the cutting edge. It is afraid that the carbon paste layer is removed easily by the change of the environment conditions.

REFERENCES

[1]   H. Tohya, N. Toya, "A Novel Decoupling Component for the Power Distribution Network", IEEE, ICGCS 2010, pp. 479-484, 2010.

[2]   H. Tohya N. Toya, "Solitary Electromagnetic Waves Generated by the Switching Mode Circuit", http://cdn.intechweb.org/pdfs/15930.pdf

Send Orders for Reprints to reprints@benthamscience.net

CHAPTER 10

Novel On-Board Decoupling Component Instead of LILC

Abstract: The novel on-board decoupling component as the substitute of LILL was developed, which became actualized by the novel characteristic equations, and the all problems of LILC were solved.

Keywords: Prototype, IEC, CISPR22, EMC, on-board decoupling component, silver coating layer, cathode, anode, carbon paste layer, conductive polymer layer, etched layer, aluminum film, low impedance lossy line (LILL), conductive polymer, magnetic field distribution, transmission coefficient, terminal impedance, branch power trace, stem power trace, void.

PROTOTYPING

Fig. **1** shows an example of the power current of the DDR2 dual-in-line memory module (DDR2 DIMM).

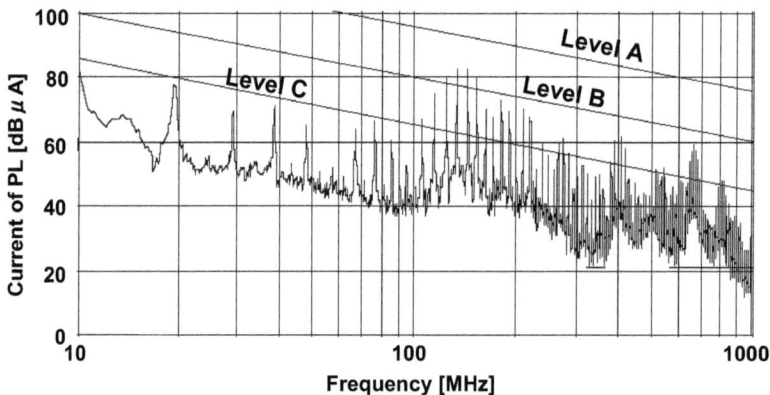

Figure 1: Power Current Spectra of DDR2 DIMM.

In Fig. **1**, the magnitude of the power current DDR2 DIMM is one of the group of the largest power current [1]. The power current of the microcomputer is large usually however the power current is divided by many power terminals. The mentioned magnetic probe which was standardized by IEC in "magnetic probe method" was used for measuring, the *x*-axis shows the frequency allocated to $1GHz$ from $10MHz$ on the log scale. The *y*-axis shows the current of the power line (PL) allocated to $100dB\mu A$ from $0dB\mu A$ on the linear scale. The peak spectrum value of the current at $134MHz$ is 81dBμA which is $11.2mA$. The power of it is $0.74mW$ because the measured input impedance was 5Ω at $134MHz$.

The electric field strength at the distance r from the antenna when EMW of P watt is radiated from the antenna is [2]:

$$E = \frac{7\sqrt{P}}{r} \ [V/m] \tag{1}$$

According to the IEC CISPR22, the Limit of the electric field at $134MHz$ from the class B information technology equipment is 37dBμV/m at the distance of 10m. From the equation 1, E is $19mV/m$ or 85.6dBμV/m when p is $0.74mW$ and the radiation efficiency is 1.0. Therefore, the attenuation of the power decoupling at the worst case should be more than 49dB at $134MHz$.

With the above as had presented, EMC will be improved greatly when the decoupling circuit is enhanced. However the decoupling performance of the capacitor is poor, and LILC and the commercialized similar components have many critical problems about the actual performance and the reliability exist in them, though they nave the high decoupling performance. In order to make a breakthrough, the development of the novel on-board decoupling component as the low impedance lossy line (LILL).

DEVELOPMENT OF NOVEL ON-BOARD LILL

Fig. **2** shows the prototypes of the on-board LILL. Fig. **2a** shows the forms of them and Fig. **2b** shows the cross-section of the chip [3, 4].

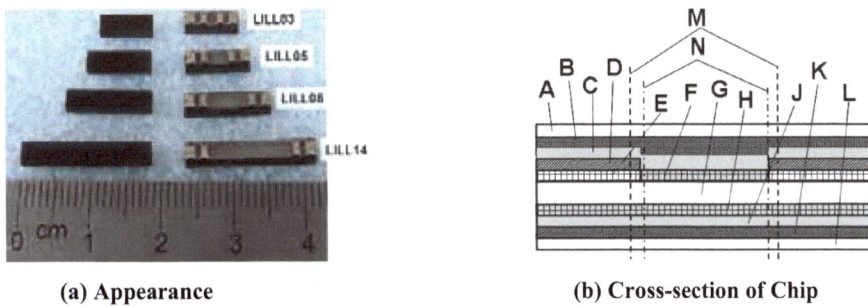

(a) Appearance   (b) Cross-section of Chip

Figure 2: Prototypes of On-board LILL.

In Fig. **2**, the chip length of each LILL03, LILL05, LILL08, and LILL14 is 3.5*mm*, 5*mm*, 8*mm*, and 14*mm*. In Fig. **2b**, the formation of the cross-section is common to four sides of the chip. A and L are the silver coating layers, A is the cathode and L is the anode, B and K are the carbon paste layers, C and J are the conductive polymer layers, E is the etched layer of the aluminum film without the conductive polymer, F and H are the etched layer of the aluminum film which was filled with the conductive polymer, and G is the aluminum layer, N is the boundary of the available chip, and M is the cutting-plane line. D is the masking layer to keep the insulation between the aluminum layer and the conductive polymer layer at the edge of the line. The conductive polymer layer is formed in the water solution of the microscopic particles of the conductive polymer. The etched layer is covered by the alumina film and the approximate thickness of the chip is 200*μm*.

Two transmission lines are formed at both side of the aluminum layer in the case of the cross section shown in Fig. **2b**. However, when the forward bias is added to the chip in the suitable period, the thickness of the alumina layer of the anode side is reduced electrochemically. In addition, the transmission line of the anode side may be shorted because the masking layer is not equipped in the transmission line of the anode side. As the result, the effective transmission line is formed at the cathode side only.

All technical problems about the decoupling capacitor and LILC were solved by the developed on-board LILL are the following;

a. The layer formation and the materials are optimized by using the novel characteristic equations and the latest technologies.

b. The thick aluminum film is unnecessary because the DC current flows in the copper film of the electrode.

c. The transmission coefficient depends on the etching surface because transmission line is formed on the etching face of the aluminum.

d. There is not the limit of the DC current because the thickness of the copper film can be set freely.

e. The transmission coefficient depends on the etching surface because transmission line is not formed at the cut-surface of the etched aluminum film.

f. The chemical reaction at the chip manufacturing becomes unnecessary by adopting the novel conductive polymer and the novel formation of the electrode.

Fig. **3** shows an example of cross-section of the LILL chip observed by the scanning electron microscope (SEM). Fig. **3a** shows the all layers. Fig. **3b** shows the close-up of the etching layer.

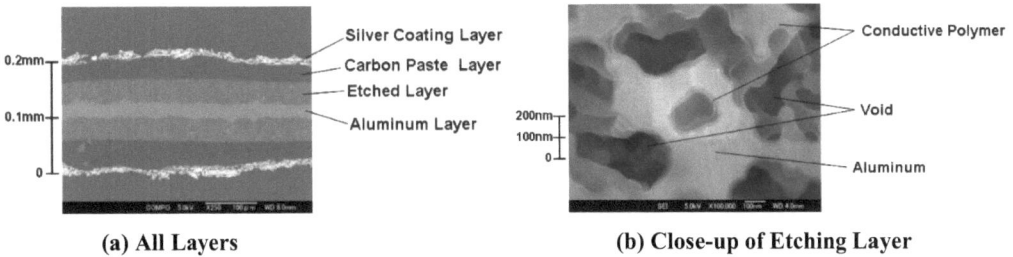

(a) All Layers   (b) Close-up of Etching Layer

Figure 3: Cross-section View of LILL Chip

In Fig. **3b,** many voids were confirmed in the etching layer. Transmission loss does not occur in the air. The size of the voids might be larger than the thickness of the alumina layer. The formation process of the conductive layer in the etching layer is based on it of the conventional capacitor. The voids are unconcerned to the performance of the capacitor because the conductive polymer and the carbon paste are permitted to be the point connection. Therefore the problem of the voids is considered to be not cleared up till LILL will get the relatively large market. In the case of LILL the void increases the terminal impedance and reduces the attenuation constant because EMW travels between the aluminum layer and the outer conductors.

Fig. **4** shows the measured magnetic field distribution of the test board at 1GHz. Fig. **5** shows the measured magnetic field distribution of the test board at 2GHz. Fig. **6** shows the measured magnetic field distribution of the test board at 3GHz. Fig. **4a**, Fig. **5a**, and Fig. **6a** show the surface. Fig. **4b**, Fig. **5b**, and Fig. **6b** show a perspective view. Fig. **4c**, Fig. **5c**, and Fig. **6c** show another perspective view. Fig. **4d**, Fig. **5d**, and Fig. **6d** show the scale of the magnetic field strength.

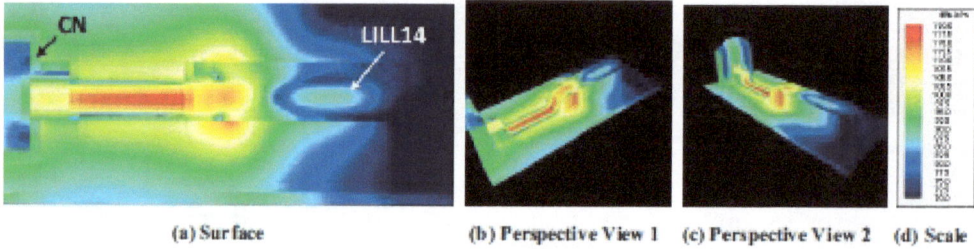

| (a) Surface | (b) Perspective View 1 | (c) Perspective View 2 | (d) Scale |

Figure 4: Measured Magnetic Field Distribution of Test Board at 1GHz.

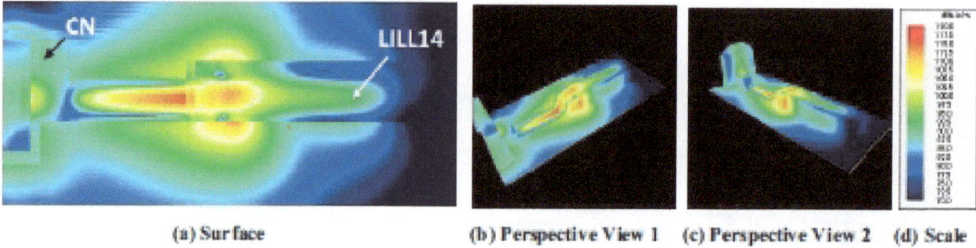

| (a) Surface | (b) Perspective View 1 | (c) Perspective View 2 | (d) Scale |

Figure 5: Measured Magnetic Field Distribution of Test Board at 2GHz

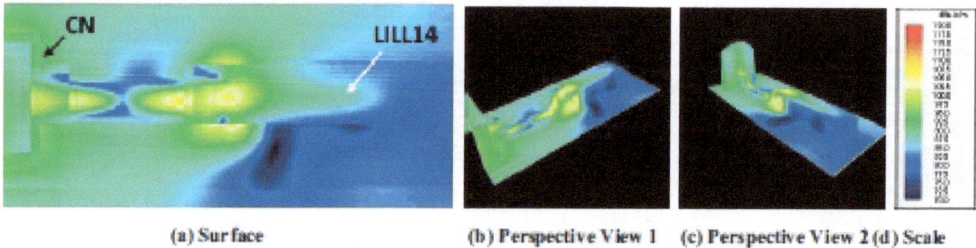

| (a) Surface | (b) Perspective View 1 | (c) Perspective View 2 (d) Scale |

Figure 6: Measured Magnetic Field Distribution of Test Board at 3GHz.

In Fig. **4**, Fig. **5**, and Fig. **6**, the test board consisted of CN which is the SMA connector, microstrip line which has the characteristic impedance of 50Ω, and LILL14. The sinusoidal signal was instilled into CN and the magnetic field was measured by the MP method. The microstrip line consisted of the strip line having the 1*mm* width and the ground plane. From Fig. **3**, Fig. **4**, and Fig. **5**, it was confirmed that the decoupling performance of LILL4 is excellent at 1*GHz*, and is good at 2*GHz* and 3*GHz*. However, the condition of the measurement is almost same to it at the network analyzer.

EVALUATION ON ACTUAL BOARD [3, 4]

Fig. **7** shows the actual board for the evaluation of the on-board LILL. Eighteen prototypes of LILL05 were used at the yellow part from LL01 to LL18.

(a) Front Surface   (b) Rear Surface

Figure 7: Actual bord for Evaluation.

In Fig. **7**, each LILL05 was placed nearby of LSI and was connected to the power receiving terminal of LSI. The test board was reworked from the actual board for the game machine, LSI01 and LSI02 was the mask ROM, LSI03 was the FPGA, LSI04 was CPU, and LSI05 was the analog IC. The oscillators of 33.25*MHz* and 66.66*MHz* were used. The size of the PCB was 156*mm*×105*mm*, and it consisted of the eight metal layers. The power trace is formed slender, and the designed average width was 10*mm* approximately.

Fig. **8** shows the trace pattern of two power distribution layers of the actual board.

Fig. **9** shows the calculated characteristics of LILL05 when they were placed on the actual board. Fig. **9a** shows the transmission coefficient and Fig. **9b** shows the terminal impedance. The red curve shows the original S_{21} which is measured by the network analyzer and the black curve shows S_{21} which is used on the test board. The difference between blue curve and the red curve at $100MHz$ is 20dB approximately.

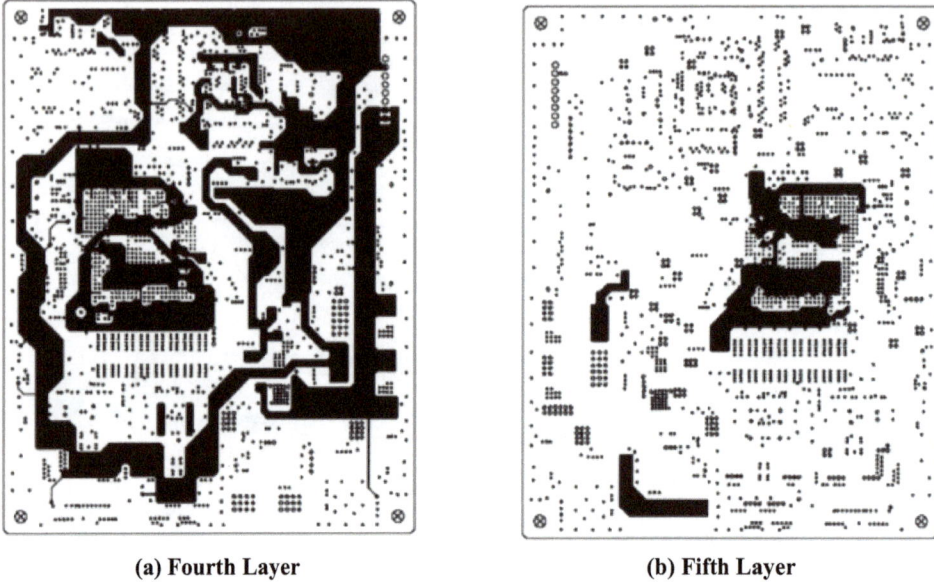

(a) Fourth Layer   (b) Fifth Layer

Figure 8: Trace Pattern of Two Power Distribution Layers of Actual Board.

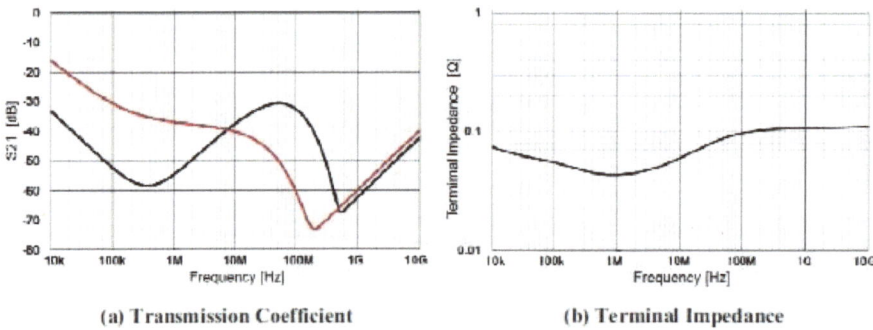

(a) Transmission Coefficient   (b) Terminal Impedance

Figure 9: Calculated Characteristics of LILL05 on Test Board.

Fig. **10** shows the measured S_{21} about FPGA (LSI03) which is shown by the blue curve by the network analyzer and the measured S_{21} which is shown by the red curve by the MP method in the actual board. Fig. **10a** shows the measured S_{21} of LL10 of LILL05. Fig. **10b** shows the measured S_{21} of LL11 of LILL05. The difference between blue curve and the red curve at 100MHz is 40dB approximately. The difference value of two curves in Fig. **10** is approximate10dB larger than it of Fig. **9**. It was supposed that this reason depend on the power traces which do not consist of the stem and the meander.

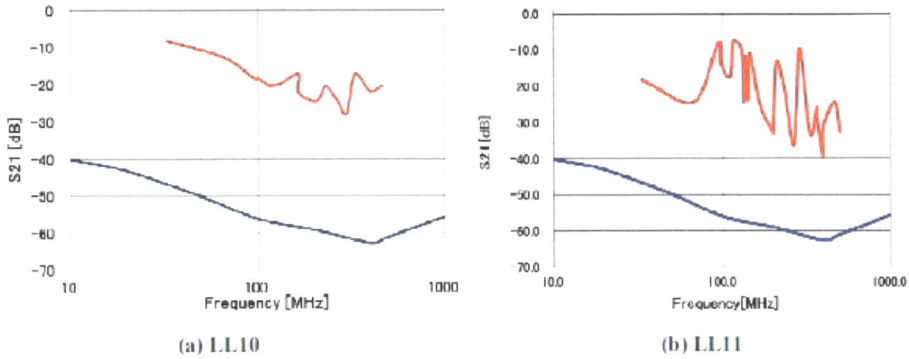

(a) LL10   (b) LL11

Figure 10: Measured S_{21} of LILL05 and it on Actual Board.

IMPROVEMENT OF ON-BOARD LILL

Fig. **11** shows the measured S_{21} of the latest prototype of the LILL14 of which rated working voltage is $10WV$. The x-axis is allocated to $3GHz$ from $100kHz$ on the log scale. The y-axis is allocated to -100dB from 0dB on the linear scale.

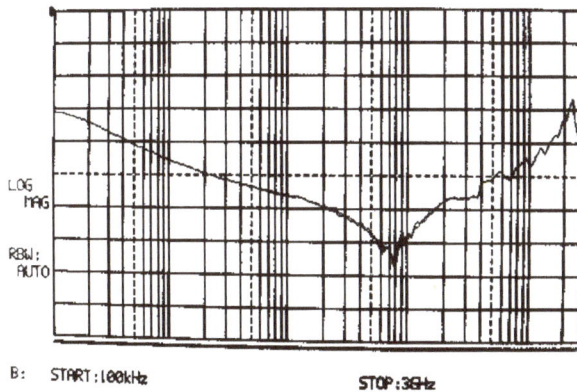

Figure 11: Example of Measured S_{21} of Latest Prototype of LILL14.

Fig. **12** shows the calculated characteristics of latest prototype of the LILL14. The calculation condition was the following; z was $14mm$, C_1 was $33.9\mu F$, R was 0.8, k was 51.3, L was 0.45, w was $1mm$, a was $22.8nm$, ρ_C was $300\mu\Omega m$, each σ of the conductive polymer and the carbon graphite was $12,000 S/m$ and $72,727 S/m$, μ_r of the alumina was 8.5, b_A for the calculation of the characteristic impedance of the attenuation constant was 0, b_A for the calculation of the characteristic impedance of the reflection coefficient was $0.8\mu m$, b_s was $1\mu m$, b_c was $2.29\mu m$, and C_T was $0.7\times10^{-16}F$. The thickness of the etched aluminum film was $110\mu m$, respectively. The calculated transmission coefficient is well matching the measured value shown in Fig. **11**.

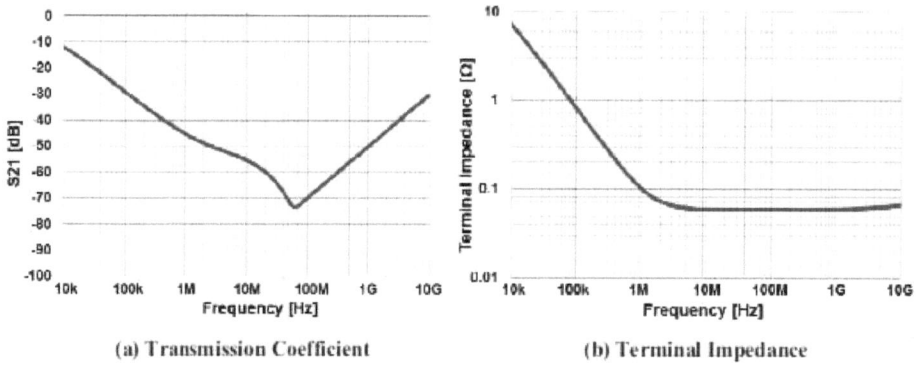

(a) Transmission Coefficient   (b) Terminal Impedance

Figure 12: Calculated Characteristics of Latest Prototype of LILL14.

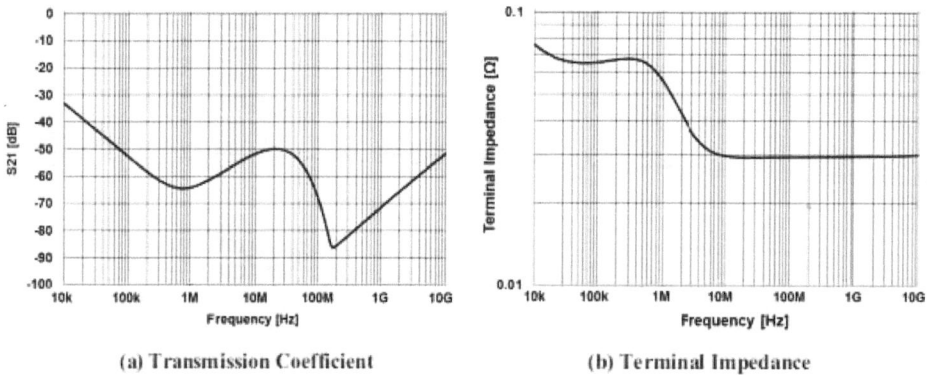

(a) Transmission Coefficient   (b) Terminal Impedance

Figure 13: Calculated On-board Characteristics of Latest Prototype of LILL14 Used on PCB.

Figure 13: Calculated On-board Characteristics of Latest Prototype of LILL14 Used on PCB.

Fig. **13** shows the calculated transmission coefficient and the terminal impedance when the latest prototype of the LILL14 is used on the PCB. The calculation condition was the following; the size of the power trace between the decoupling component and LSI was $40mm\times40mm$ and the size of the power trace between the power source and the decoupling component was $10mm \times 200mm$, the thickness the insulator layer between the power traces and the ground plane is $0.1075mm$ and its dielectric constant is 4.35, and the conventional capacitor of $1mF$ was connected in parallel, respectively.

In Fig. **13a**, the transmission coefficient is -69dB at $134MHz$ though the transmission coefficient is degraded at the frequency band to $100MHz$ from $10MHz$ approximately. Therefore the attenuation is large enough than the above mentioned desirable performance of LILL even if it is used on actual PCB.

Fig. **14** shows the calculated characteristics of the modified latest prototype LILL of 4WV, and the parameters are same as the latest prototype of 14.8WV except *a* is $6.168nm$.

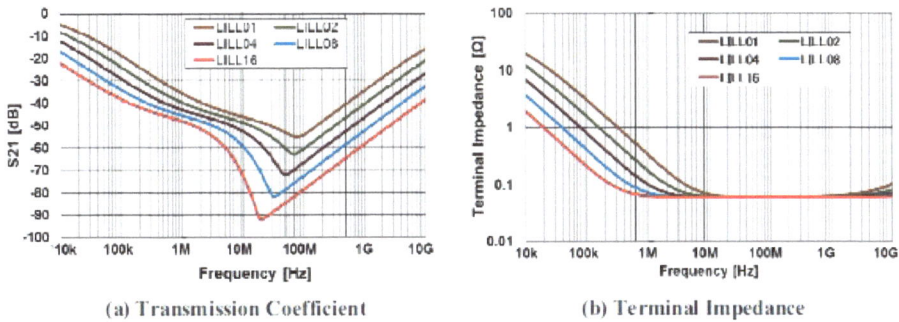

(a) Transmission Coefficient   (b) Terminal Impedance

Figure 14: Calculated Characteristics of Modified Latest Prototype of LILL.

In Fig. **14**, the calculated transmission coefficient is being improved.

Fig. **15** shows the calculated transmission coefficient and the terminal impedance when the modified latest prototype of the LILL is used on the PCB. In this case, PDN consists of the branch power traces and the stem power trace similar to the the redesigned pattern of the PDN layer of the MB of the first workstation in Chapter 4.

In Fig. **15**, the calculation condition was the following; LILL is connected to the branch power traces of $1mm \times 10mm$, which is connected to stem power trace of $200mm$ width, and the others are same as the calculation for Fig. **13**. In Fig. **13a**, the transmission coefficient is -69dB at $134MHz$ though the transmission coefficient is degraded at the frequency band to $100MHz$ from $10MHz$

approximately. Therefore the attenuation is large enough than the above mentioned desirable performance of LILL even if it is used on actual PCB.

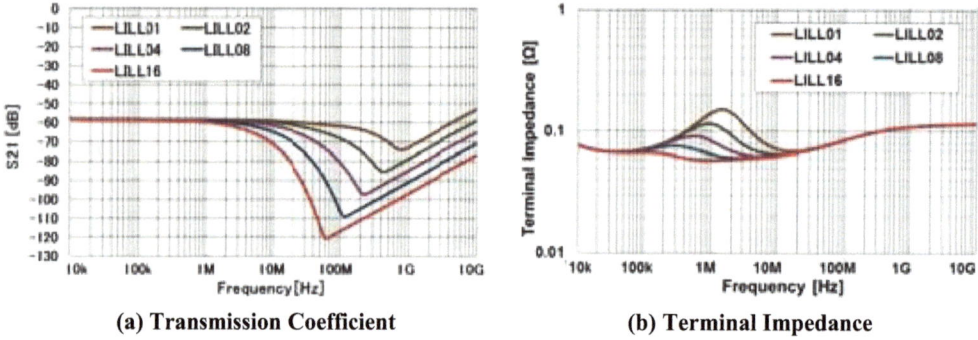

(a) Transmission Coefficient   (b) Terminal Impedance

Figure 15: Calculated On-board Characteristics of Modified Latest Prototype of LILL on PCB.

When the problem of the voids is solved, the more improved the characteristics of the LILL was hoped. Fig. **16** shows the calculated transmission coefficient and the terminal impedance of the example of the desirable LILL.

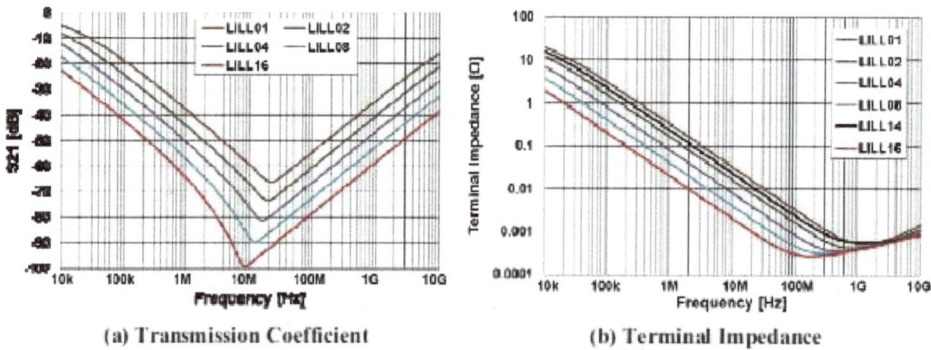

(a) Transmission Coefficient   (b) Terminal Impedance

Figure 16: Calculated Characteristics of Example of Desirable LILL.

In Fig. **16**, the chip of the desirable LILL is being improved, the calculation condition was the following; z was to $16mm$ from $1mm$, C_1 was $10\mu F$, R was 1.0, k was 15.14, L was 1.0, b_A was 0, b_c was $1.38\mu m$, respectively, and other parameters were same as they of the latest prototype of the LILL14.

Fig. **17** shows the calculated transmission coefficient and the terminal impedance when the example of the desirable LILL is used on the PCB. The calculation condition is same as it of the latest prototype of the LILL14.

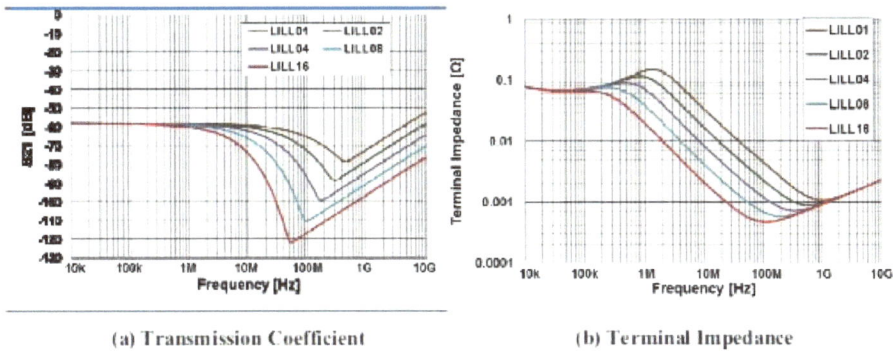

(a) Transmission Coefficient   (b) Terminal Impedance

Figure 17: Calculated Characteristics of Example of Desirable LILL Used on PCB.

In Fig. **17**, the calculation condition is same as it for Fig. **15**. In Fig. **17a**, the transmission coefficient is relatively smaller than it of Fig. **15a**, however the terminal impedance is being greatly improved at the frequency higher than several megahertz.

CONCLUSIONS

The performance of the conventional decoupling method using the capacitors is not suitable to form QSCC. The developed LILL has excellent features. The developed LILL can suppress EMI effectively when LILL is connected directly to the power supply terminal of LSI or the transistor and when it connected to the power receiving terminal of PCB. The characteristic impedance of the power line connected to LILL is desired to be relatively high. The improvement of the current process to form the conductive polymer layer of LILL will be effective to get more excellent performance.

LILL is useful for forming QSCC of PDN. However the LILL technology cannot be applied to the on-chip interconnect in accordance with the conventional theory because the digital voltage on the chip consists of many harmonic waves which are similar to them existing on the board. In other words, the size of the LILL for applying to PCB and LSI will become same when the same decoupling performance is required. Generally, the solution technique becomes simpler whenever more approaching to the trouble source. From this, above idea is considered to be doubtful. Therefore, this idea based on the AC circuit theory adopting the Fourier Transform was needed the discussion.

REFERENCES

[1]   H. Tohya, T. Shimasaki, K. Mori, T. Osato, N. Kuwabara, H. Muramatsu, "Simple and Convenient EMI Management Method for Modules of IT Equipment", IEEE, EMC EUROPE 2012, P-3-5-1, 2012.

[2]   IEC, "Specification for radio disturbance and immunity measuring apparatus and methods", CISPR 16-2-3, pp. 79, 2003.

[3]   H. Tohya, N. Toya, "A Novel Decoupling Component for the Power Distribution Network", IEEE, ICGCS 2010, pp. 479-484, 2010.

[4]   H. Tohya, N. Toya, "Solitary Electromagnetic Waves Generated by the Switching Mode Circuit", http://cdn.intechweb.org/pdfs/15930.pdf

Send Orders for Reprints to reprints@benthamscience.net
Switching Mode Circuit Analysis and Design, 2013, 145-156   **145**

CHAPTER 11

Advocating SEMW Theory [1- 4]

Abstract: The SEMW theory was developed for SMC design and analysis by fusing the conventional semiconductor physics, conventional nonlinear undulation theory and the conventional EMW theory.

Keywords: Nonlinear undulation theory, solitary wave, great wave, PMOSFET, NMOSFET, SMC, SEMW, on-chip inverter, semiconductor physics, saturation threshold voltage, drain current, gate voltage, drain source voltage, soliton, solitary electric field wave (SEW), significant frequency (SF), modified significant frequency (MSF), vector wave equations, solitary magnetic field wave (SMW), electric current.

GENERATION MECHANISM OF SEMW IN SMC

The nonlinear undulation theory had been established in the physics, which includes the solitary wave. It was discovered as the great wave by John Scott Russell at the experiment at the canal. Intuitively, it was considered that the canal corresponds to the transmission line and the gate of canal corresponds to the switching transistor in the SMC.

LSI is the typical circuit of the SMC. Fig. **1** shows the equivalent circuit of the on-chip inverter which forms the elementary circuit of the LSI. The elementary circuit of the LSI consists of the power supply PS, the on-chip inverter Z_1 and Z_2, the power line, and the signal line. Z_1 consists of the complementary circuit consisting of PMOSFET Q_1 and NMOSFET Q_2. The gate circuit which is the fundamental circuit of LSI consists of several on-chip inverters. The ON state of the on-chip inverter means the state that Q_1 is ON and Q_2 is OFF. The OFF state of the on-chip inverter means the state that Q_1 is OFF and Q_2 is ON. The power supply PS supplies V_{DD}. PMOSFET Q_1 governs the on-chip inverter Z_1 because PMOSFET is slower than NMOSFET. The drain current at the period of switching ON of the on-chip inverter Z_1 will consist of the symmetrical waveform on the time axis because the on-chip inverter consists of the complimentary circuit, which is governed by PMOSFET. The drain current increases at the first half period of the switching ON and it decreases at the last half period.

Hirokazu Tohya

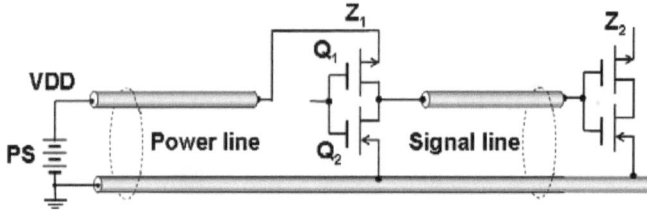

Figure 1: Elemantary Circuit of LSI.

According to the semiconductor physics and the ITRS, the drain current of the PMOSFET higher than the saturation threshold voltage at the first period is:

$$I_{d1}(V_g) = A \cdot I_{dsp} \ (V_g - V_{TS})/z(V_{dd} - V_{TS}) \tag{1}$$

where A is the conversion coefficient, V_{dd} is the supply voltage, I_{dsp} is the peak saturation drain current of PMOSFET, V_{Ts} is the saturation threshold voltage, z is the gate width, and V_g is the gate voltage.

The drain current of the PMOSFET up to the saturation threshold voltage at the first period is:

$$I_{d2}(V_g) = \frac{I_{dsl}}{z} \cdot \frac{V_g}{V_{dd}} \tag{2}$$

where I_{dsl} is the sub threshold leakage current of the drain of PMOSFET.

The summation drain current of PMOSFET at the first period is:

$$I_d(V_g) \equiv I_{d1}(V_g) + I_{d2}(V_g) \tag{3}$$

The drain current of PMOSFET at the last period is:

$$I_{dR}(V_g) \equiv I_d(1.1 - V_g) \tag{4}$$

The drain current of the on-chip inverter corresponding to the gate voltage at the period of switching ON of the on-chip inverter Z_1 is:

$$I_{dI}(V_g) = \left(I_d^{-n}(V_g) + I_{dR}(V_g)^{-n}\right)^{\frac{-1}{n}} \tag{5}$$

Fig. **2** shows the calculated drain current of the on-chip inverter corresponding to the gate voltage. In Fig. **2**, the calculation condition which corresponds to the 2009 technology node of the ITRS 2008 update was the following; V_{dd} was $1.1V$, I_{dsp} was $1.317mA/\mu m$ at V_{dd}, V_{Ts} was $196mV$, I_{dsl} was $0.17\mu A/\mu m$ at V_{dd}, was $108nm$, and n in the equation 5 was 30, respectively. In Fig. **2**, the calculation condition which corresponds to the 2009 technology node of the ITRS 2008 update was the following; V_{dd} was $1.1V$, I_{dsp} was $1.317mA/\mu m$ at V_{dd}, V_{Ts} was $196mV$, I_{dsl} was $0.17\mu A/\mu m$ at V_{dd}, was $108nm$, and n in the equation 5 was 30, respectively.

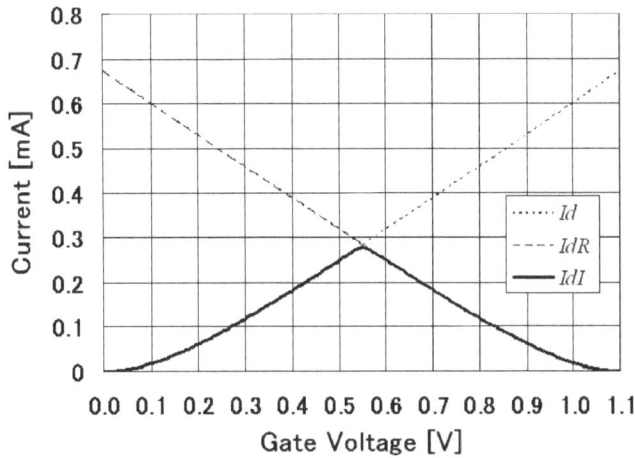

Figure 2: Calculated Drain Current Corresponding to Gate Voltage of On-chip Inverter.

According to the ITRS, the voltage between drain and source of PMOSFET is:

$$V_{ds}(V_g) = \frac{1}{C_{ox}} \int_0^{2T_1} I_d(V_g)dt \qquad (6)$$

where C_{ox} is the gate capacitance.

When V_{ds} is $0.55V$ and C_{ox} is $0.875fF$ which is in accordance with 2009TN of the ITRS 2008 update, T_1 is $1.7ps$ from the equation 6.

Fig. **3** shows the calculated drain source voltage of the on-chip inverter. The x-axis is being converted to the time from the gate voltage, and $0ps$ corresponds to $0V$ and $3.4ps$ corresponds to $1.1V$.

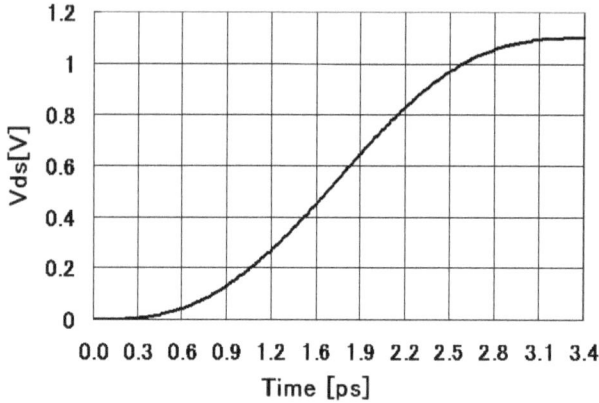

Figure 3: Calculated Drain Source Voltage of On-chip Inverter.

Fig. **4** shows the calculated I_{d2}, I_{dR2} and I_{dl2} of the second stage on-chip inverter.

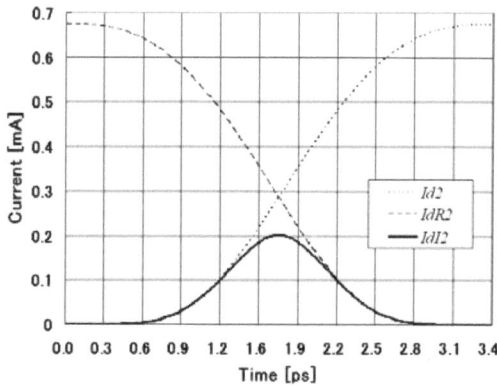

Figure 4: Calculated Drain Current of Second Stage On-chip Inverter.

In Fig. **4**, the drain current was calculated by the equations of 3 to 5 by adopting the drain source voltage shown in Fig. **3** as V_g.

According to EMW theory, the time-domain waveform of the magnetic field is same as the current waveform except the magnitude. The magnitude of the magnetic field depends on the physical parameter of each on-chip inverters.

Fig. **5** shows the calculated relative magnetic field of the second stage on-chip inverter which corresponds to the waveform of the drain current shown 4. 1.0 of the y-axis corresponds to the peak value of the drain current shown Fig. **4**.

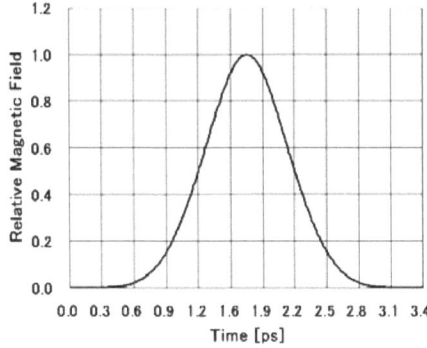

Figure 5: Calculated Magnetic Field of Second Stage On-chip Inverter.

Fig. **6** shows the calculated drain source voltage of the second stage of the on-chip inverter. The drain source voltage was calculated by the equations of 3 to 5 by adopting the drain source voltage shown in Fig. **3** as V_g, the rise time is 0.98*ps*.

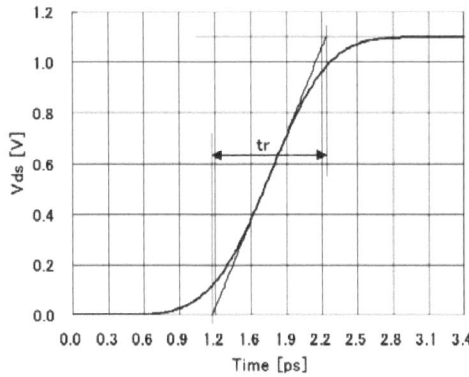

Figure 6: Calculated Relative Magnetic Field of Second Stage On-chip Inverter.

Similarly, the relative electric field between drain and source (E_{ds}) of the on-chip inverter can be got by the total sum of the increasing electric field of the first half of the switching period and the decreasing electric field of the last period.

E_{ds} of the first half of the switching period is:

$$E_{ds}(V_g) = \frac{1}{C_{ox}} \int_0^{2T_1} I_d(V_g) dt \tag{7}$$

E_{ds} of the last half of the switching period is:

$$E_{dsR}(V_g) = \frac{1}{C_{ox}} \int_0^{2T_1} I_d(1.1 - V_g)dt \tag{8}$$

The electric field between drain and source of the on-chip inverter is:

$$E_{dI}(V_g) = \left(E_d^{-n}(V_g) + E_{dR}(V_g)^{-n}\right)^{\frac{-1}{n}} \tag{9}$$

Fig. **7** shows the calculated relative electric field of the second stage of the on-chip inverter. In Fig. **7**, the rise time corresponds to it shown in Figure. The drain current and the drain source voltage about the third stage of the on-chip inverter were calculated. However it did not finish at one week on our relatively high-performance desk top computer. The calculated waveform of the drain current of the third stage of the on-chip inverter is estimated to be slimmer than it shown in Fig. **4**.

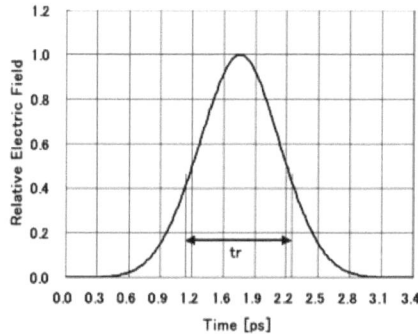

Figure 7: Calculated Relative Electric Field of Second Stage On-chip Inverter.

According to the electromagnetism, each varying magnetic field and electric field has same shape and forms EMW. Therefore, the magnetic field shown in Fig. **5** and the electric field shown in Fig. **7** will form equally EMW. In addition, each waveform is same and has the similar shape of the soliton. From above, EMW which is generated by the on-chip inverter was named the solitary electromagnetic wave (SEMW).

SEMW is a kind of the soliton. The wave equation of the soliton which has not the mass in accordance with the nonlinear undulation theory is:

$$u(t) = A \cdot \sec^2 h\left(B(t - T_1)\right) \tag{10}$$

where A is the magnitude, B is the waveform constant, T_1 is the initial time.

Fig. **8** shows the solitary electric field wave (SEW) calculated by the equation 10.

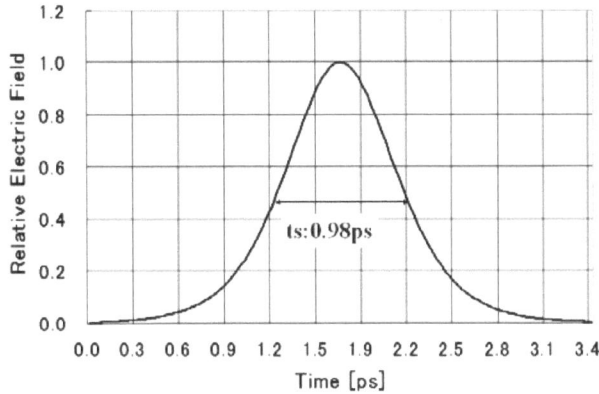

Figure 8: Calculated Relative Electric Field of Second Stage On-chip Inverter.

In Fig. **8**, the calculation condition was the following; $T_1 = 1.7ps$, $A = 1$, $B = 20.55$, t was set to 3.4ps from 3.7fs, respectively. t_s is defined as the effective interval and it equals to the rise time of the on-chip inverter, and the traveling distance during t_s is defined as the wave length of SEMW.

The wave length of the SEMW is:

$$\lambda_S = \frac{t_s}{\sqrt{\mu_0 \varepsilon_r \varepsilon_0}}$$

(11)

When the dielectric constant (κ or ε_r) of the on-chip interconnect is 2.7 which is got from the 2009TN of the ITRS 2008 update, 0.98ps of the effective interval t_s in Fig. **8** corresponds to 179μm of the wave length.

Fig. **9** shows the rising part of the signal voltage of the on-chip inverter. The maximum magnitude was set to 1.1V, and the shape of the rising part of the signal voltage is similar to the wave form of V_C which consists of the sinusoidal wave. The calculation for getting the waveforms of Fig. **8** and Fig. **9** was finished in a moment on the typical desktop computer. This means that the analysis and design of the SMC based on the SEMW theory become easy.

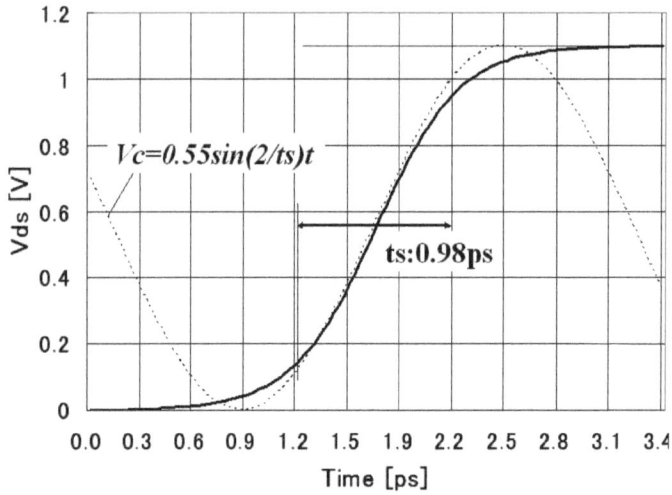

Figure 9: Rising Part of Signal Voltage of On-chip Inverter.

The definition of the significant frequency (SF) had been presented [5]. This definition was considered to be true for $7 \leq r \leq 13$, where $r \equiv T/tr$, T is the pulse width, and t_r is the ramp time of the pulse. The outline of this definition is the following; SF of a trapezoidal pulse is defined as $0.34/t_r$, the first property of SF is that about 15% of its frequency components are at frequencies higher than SF for each trapezoidal pulse, and the second property of SF is that the magnitude of pulse's spectrum at a frequency higher than the SF is less than 10% of its maximum value and the spectral amplitude rolls off much faster than 20dB/decade beyond the SF.

This definition cannot be directly applied to the SEMW because the SEMW has not harmonic waves. However, above properties of the SF can be used for the similarity of the wave shapes between the soliton and the sine wave in the Fig. **9**. SF of $0.34/t_s$ is similar to $1/\pi t_s$. From above, the SF was redefined as the modified significant frequency (MSF) which is $1/\pi t_s$ in the SEMW theory.

The properties of the MSF in accordance with the SF are the following;

a.   The correlation of the time domain wave shape of the SEMW and half wave shape of the sine-wave having the frequency of $1/\pi t_s$ is more than 85%.

b.   No spectra exist at the frequency band higher than MSF.

MSF of $0.98ps$ which is the gate delay of 2009TN shown in the ITRS 2008 update is $325GHz$. MSF can be used for to design and analysis of the component and the transmission line on the SMC. This way using MSF will be effective enough until the simulators and the measurement instruments are developed by applying the SEMW theory.

The developed vector wave equations of the SEMW are

$$\dot{E}(t) = i\sqrt{2}E_0 \cdot \sec^2 h \left(B\left(t \mp z\sqrt{\mu_0\varepsilon_0} \right) + T_1 \right) \tag{12}$$

$$\dot{H}(t) = \pm j\sqrt{2} \sqrt{\frac{\varepsilon_0}{\mu_0}} E_0 \cdot \sec^2 h \left(B\left(t \mp z\sqrt{\mu_0\varepsilon_0} \right) + T_1 \right) \tag{13}$$

The equations of 12 and 13 were the development fruits which are gotten from fusing the conventional nonlinear undulation theory and the conventional EMW theory.

BEHAVIOURS OF SEMW ON TRANSMISSION LINE

According to the developed vector wave equations of the SEMW, the SEMW consists of SEW and the solitary magnetic field wave (SMW). These waves are at right angles to each other and both wave shapes are same except its magnitude. The SEMW travels in the insulator of the transmission line at $1/\sqrt{\mu\varepsilon}$ of the light speed as EMW.

The normal SMC is a kind of the voltage source circuit. In this case, SEW acts as the leading player, and the SMW depends on the SEW.

Fig. **10** shows the elementary SMC.

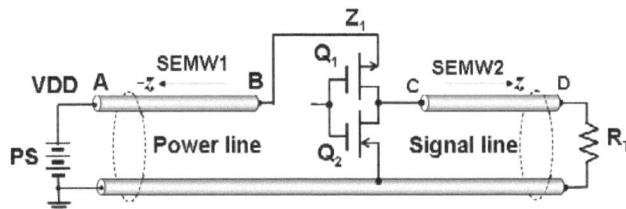

Figure 10: Elementary SMC.

In Fig. **10**, SEMW1 and SEMW2 are generated by the on-chip inverter (Z1). According to the electromagnetism, the law of conservation of energy, and the equation 12 and 13, SEMW1 has the positive magnitude and SEMW2 has the negative magnitude. SEMW1 travels on the power line toward the power supply with feeding back the electrostatic energy to the power supply. SEMW2 travels on the signal line to the terminal D with pulling out the electrostatic energy from the power source to the signal line.

SEMW1 consists of SEW1 and SMW1. SEMW2 consists of SEW2 and SMW2. In the voltage source SMC, the SEW governs SMC usually and the magnitude of SMW depends on the source voltage and the characteristic impedance of the signal line and the power line.

Fig. **11** shows the wave shapes on the transmission line connected to the on-chip inverter. Fig. **11a** shows SEW1 and the falling part of the voltage on the power line (V_{power}). Fig. **11b** shows SEW2 and the rising part of the voltage on the signal line (V_{signal}).

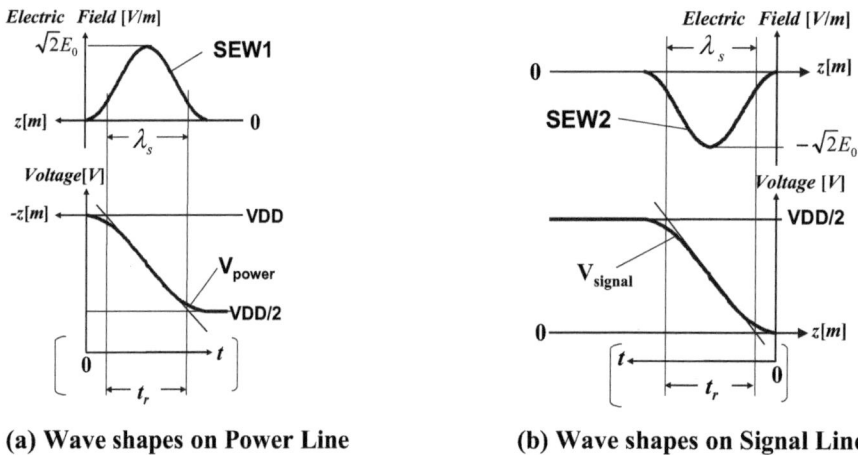

(a) **Wave shapes on Power Line**   (b) **Wave shapes on Signal Line**

Figure 11: Wave Shapes on Transmission Line Connected to On-chip Inverter.

In Fig. **11a**, the power line is discharged to VDD/2 [V] from VDD [V] by the traveling SEW1. In Fig. **11b**, the signal line is charged to VDD/2 [V] from zero [V] by the traveling SEW2. When the power line is discharged, the electrostatic energy on the power line is consumed by the on-chip inverter. The electrostatic energy is pulled out from the power supply when the signal line is charged. When the electrostatic energy is pulled out, the displacement current which is in accordance with the Maxwell's equations is observed. This displacement current exists during charging the transmission line. On the other hand the charge current flows in the terminal resistor after the signal line is charged.

Three kinds of the electric current on SMC were clarified by the SEMW theory are the following;

a. The first electric current is the value of the line integral of SMW around the conductor in accordance with the Ampère's circuit law, and it travels at sub light speed.

b. The second electric current is the value of the line integral of the electrostatic field around the conductor in accordance with the Ampère's circuit law.

c. The third electric current is the charge current in the conductor, which flows when the resistor is connected to the transmission line. The average drift speed is on the order of a millimeter per second.

The circuit analysis and design of SMC will become accurate and reliable by recognizing these three kinds of the electric current. These three kinds of the electric current should exist in all kind of the AC circuit. However the three kinds of the electric current are difficult to recognize on the analog circuit because EMW is vibrating continuously.

REFERENCES

[1] H. Tohya, N. Toya, "A Novel On-Chip Power Distribution Network Circuit Technology being able to actualize High-speed Signal Transmission and reduce EMI", EIC/ KEES, 2006 Korea-Japan Joint Conference on AP/EMCJ/EMT, pp. 205-208, 2006.

[2] H. Tohya, N. Toya, "DESIGN METHODOLOGY OF ON-CHIP POWER DISTRIBUTION NETWORK" IEEE, Fifth IEEE Dallas Circuit and System Workshop, pp.79-82, 2006.

[3] H. Tohya, N. Toya, "A Novel Design Methodology of the On-Chip Power Distribution Network Enhancing the Performance and Suppressing EMI of the SoC", IEEE, ISCAS 2007, pp. 889-892, 2007.

[4]   H. Tohya, N. Toya, "Solitary Electromagnetic Waves Generated by the Switching Mode Circuit", http://cdn.intechweb.org/pdfs/15930.pdf

[5]   H.B. Bakoglu, "Circuits, Interconnections, and Packaging for VLSI", pp. 239-244, Addison-Wesley Pub., 1990.

Send Orders for Reprints to reprints@benthamscience.net
Switching Mode Circuit Analysis and Design, 2013, 157-166 157

CHAPTER 12

Novel On-Chip LILL Technology [1- 5]

Abstract: The on-chip LILL technology became possible to actualize by the SEMW theory. The on-chip LILL will shorten the rise time and improve EMC of LSI greatly.

Keywords: On-chip LILL, rise time, fall time, delay time, radiated emission, gate delay, terminal impedance, power line, voltage vibration, high-speed transmission line, signal voltage, oscillator, driver, receiver, characteristic impedance, on-chip interconnect, n-type semiconductor layer, p-type semiconductor layer, on-chip interconnect, on-chip inverter.

INFLUENCE OF RISE TIME OF SIGNAL BY POWER LINE LENGTH

Fig. **1** shows the elementary SMC used with LILL. The LILL is connected between the power supply and the elementary SMC.

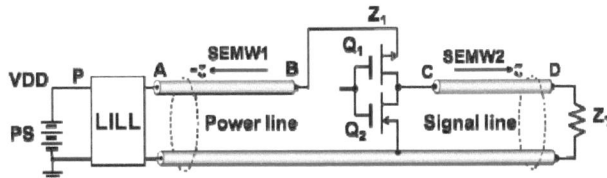

Figure 1: Elementary SMC used with LILL.

Fig. **2** shows the calculated wave shapes of the SEW and the rising part of the signal voltage at the point D in Fig. **1** when the on-chip inverter changes to ON.

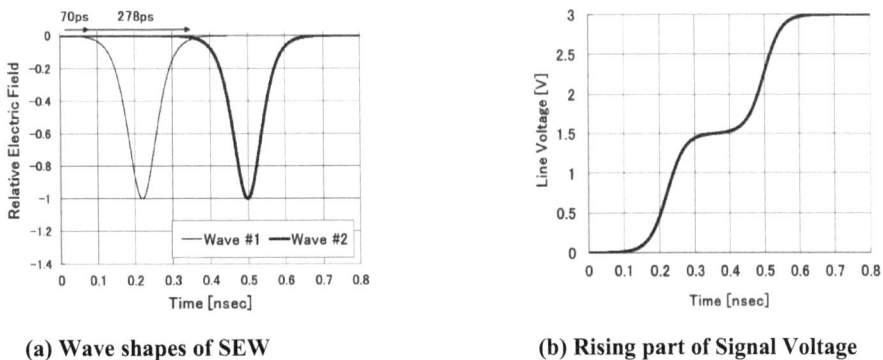

(a) **Wave shapes of SEW**　　　　(b) **Rising part of Signal Voltage**

Figure 2: Wave Shapes on Transmission Line Connected to On-chip Inverter.

Hirokazu Tohya

In Fig. **2a**, the wave #1 is SEW which was generated on the signal line, the wave #2 is SEW which was generated on the power line and it was reflected at the point A because the terminal impedance is supposed to be zero. The calculated delay time of the round trip between the point A and the point B is 278*ps*, and the calculated delay time between the point C and the point D is 70*ps*. In Fig. **2b**, the rising part of the signal voltage at the point D was got by integral of the waveforms in Fig. **2** and the magnitude of the signal voltage was set to 3V.

From Fig. **2**, it was clarified that the rise time does not only depend on gate delay of the on-chip inverter but also depends on the length of the power line from the on-chip inverter to the terminal having the low impedance, the wave shapes were simulated in accordance with the SEMW theory. The calculation condition is the following; the gate delay of the inverter is 100*ps*, ε_r is 4.35, each length of the signal line and the power line is 10*mm* and 20*mm*, and the terminal resistor RT has the impedance which matches to the characteristic impedance of the signal line.

Fig. **3** shows the calculated wave shapes of the SEW and the repeating signal voltage at the point D in Fig. **1**.

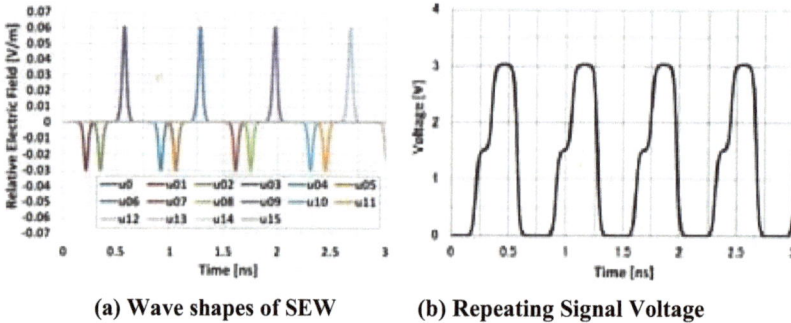

(a) **Wave shapes of SEW**   (b) **Repeating Signal Voltage**

Figure 3: Wave Shapes on Transmission Line Connected to On-chip Inverter.

In Fig. **3**, the calculation condition was the following; the gate delay of the on-chip inverter is 100*ps*, ε_r is 4.35, each length of the signal line and the power line is 10*mm* and 20*mm*, and the terminal resistor R_T has the impedance which matches to the characteristic impedance of the signal line. The resistors for the termination are not used usually on the on-chip interconnect for reducing the power consumption.

Fig. **4** shows the simulated repeating signal voltage at the point D in Fig. **1** when the power line length is changed and the termination resistor R_T is not connected. In Fig. **4**, it was also clarified that the rise time increases in proportion to the length of the power line, however, the voltage on the power line is vibrating after the on-chip inverter changes to OFF. The signal voltage was simulated by ApsimRLGC® and HSPICE. The simulation result was considered to be reliable because the loss-less transmission line is used in the sub-circuit. Fig. **1** was used as the equivalent circuit for the simulation. The simulation condition was the following; VDD was 2.5V, the value of the terminal resistor R_T was 51Ω, the terminal impedance of the power supply is zero, and the on-chip inverter (Z1) was replaced to the driver which was described by the sub-circuit depending on the design parameter, the characteristic impedance of the signal line and the power line of the actual PCB was 51.28Ω, respectively.

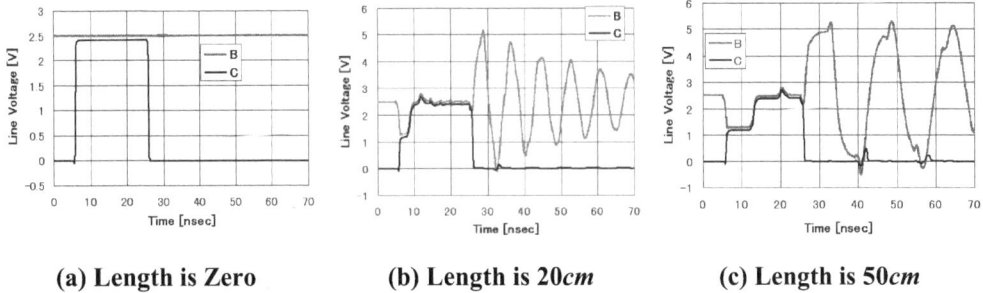

| (a) Length is Zero | (b) Length is 20*cm* | (c) Length is 50*cm* |

Figure 4: Simulated Signal Voltage When Power Line Length is Changed.

ANALYSIS OF MECHANISM OF VOLTAGE VIBRATION

The mechanism of the voltage vibration of the power line is considered to be the following;

a. The charge current exists in the conductor of the power line before changing OFF of the driver.

b. SMW is generated at the point B on the power line when the driver changes OFF but SEMW is not generated at the point C on the signal line because the point C is shorted to the ground plane when the driver is changes OFF.

c.   SEW is generated simultaneously with the generation of SMW on the power line and SEMW is formed by them.

d.   SEMW on the power line is coming and going on the power line, because the terminal-impedance of the point A is very small and it of the point B is very large.

Fig. **5a** shows the calculated waveform of SEW and Fig. **5b** shows the voltage shape, which are at the point B on the power line when the driver changes OFF. In Fig. **5a**, u01 is the generated SEW when the driver changes OFF, u02 is the 1st reflected SEW at the point B, u03 is the 2nd reflected SEW at the point B, u04 SEW is the 3rd reflected SEW at the point B, and so on. The calculation condition was the following; the rise/fall time of the driver was 100*ps*, ε_r was 4.35, and the length of the power line was 200*mm*. In Fig. **5b**, the calculated voltage is similar to the rectangular wave because 2*kΩ* at MSF was used as the terminal impedance of the off state of the driver for the simplified calculation. The voltage shape is irrelevant to EMI because it is got by the integral of the SEW. EMI is caused by the cyclic generation of SEMW on the power line. In Fig. **5**, 360*MHz* which is the repetition frequency of the even or odd number of reflected SEW and 720*MHz* which is the repetition frequency of the sum total reflected SEW will be measured as the radiated emission. The effective methods for suppressing the vibration are to shorten the power line to the terminal of the zero impedance from the driver, or to use no terminal resistors. The terminal resistor has not been used on the chip of the LSI usually. However, the terminal resistors have been used to the high-speed transmission line on the board usually. Therefore, the voltage of the on-chip power line of the driver never vibrates by this cause.

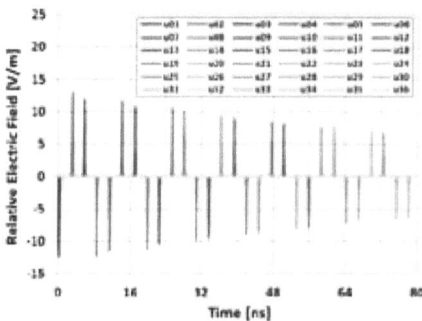

| (a) SEW Waveform | (b) Voltage Shape |

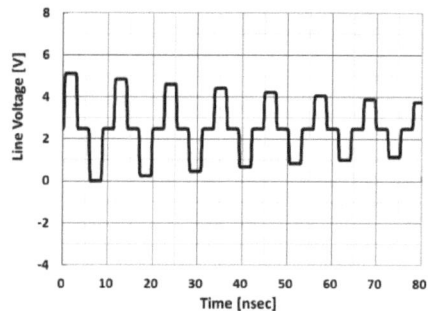

Figure 5: Waveform of SEW and Voltage Shape at Point B on Power Line.

EXPERIMENT

Fig. **6** shows the setup of the test board for confirming the relationship between power line and rise time of the signal voltage by the experiment.

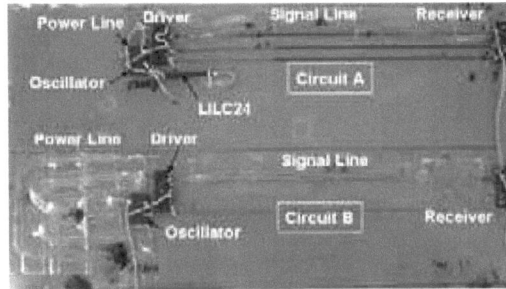

Figure 6: Setup of the Test Board.

In Fig. **6**, CMX309HWC50MHz was used for the oscillator, SN74AS00N was used for the driver and the receiver, and the prototyped LILC24 is used to reduce the impedance of the power line of the circuit A and B and to shorten the length of the power line of the circuit A. The rated typical rise time of SN74AS00N is 0.5*ns*. According to the idea of the MSF, 0.5*ns* corresponds to 637*MHz*. Each length of power line of the circuit A and circuit B is 1cm and 36*cm*. The characteristic impedance of all transmission lines is designed to be 73Ω. The resistors for impedance matching are not being used at terminal of the signal line. The polyurethane enamelled wire was used for the signal line and power line on the unprocessed single-sided PCB of FR-4.

Fig. **7** shows the measured signal voltage at the receiver.

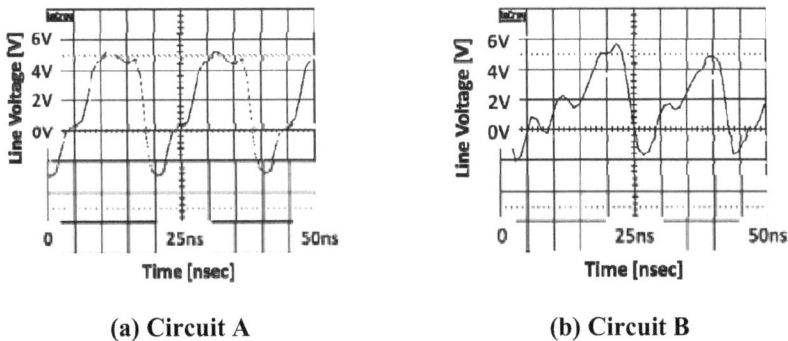

(a) Circuit A   (b) Circuit B

Figure 7: Measured Signal Voltage at Receiver.

In Fig. **7a**, the rise time of the signal voltage at the receiver of the circuit A is $3ns$ approximately. In Fig. **7b**, the rise time of the signal voltage at the receiver of the circuit B is $8ns$ approximately. Each calculated round-trip time of the length of the power line of $36cm$ and $1cm$ is $5ns$ and $0.14ns$.

As the result, SEMW theory clarified the relationship between power line and rise time of the signal voltage. When the LILL is located nearby of the on-chip inverter, it is expected that the rise time of the signal will be greatly reduced and it will become to the gate delay of the on-chip inverter, and the vibration on the power line will be suppressed. It is one of the great advantages of the on-chip LILL. To suppress EMI caused by the power line on the chip is another advantage.

Feasibility of On-Chip LILL
If the size of LILL is minimized to the microscopic size and has the desirable performance comparable to the on-board LILL, The on-chip LILL will be actualized.

According to the conventional circuit theory depending on the Fourier transform, the chip length should be same to it of the on-board LILL because the frequency of the harmonic waves on both PCB and LSI chip is not different. On the other hand, if the SEMW theory is reliable, minimizing of the size of LILL on the LSI chip will become possible. The SEMW is almost filling the power line nearby the on-chip inverter and the MSF of the SEMW generated by the on-chip inverter is $325GHz$ at 2009TN shown in the ITRS 2008 update for example.

The performances of the on-chip LILL depend on the characteristic impedance of the on-chip interconnect. The desirable terminal impedance of the on-chip LILL is decided by the relationship to the characteristic impedance of the on-chip interconnect, and its value will be up to several percent of the characteristic impedance of the on-chip interconnect including the power supply wiring layers. The on-chip interconnect consists of single wire and the insulator. However it can be considered to be handled as the transmission line which is formed by a wire, insulator layer, and the back-surface or substrate which has relatively low resistance because the distance between the interconnect and the back-surface or substrate. The characteristic impedance was calculated by ApsimRLGC® which was developed and commercialized for analysis of the transmission line, MW STUDIO® which was developed and commercialized for analysis of the wave guide, and the commonly-used equation of the microstrip line.

Fig. **8** shows the models of the on-chip interconnect for the simulation and the calculation.

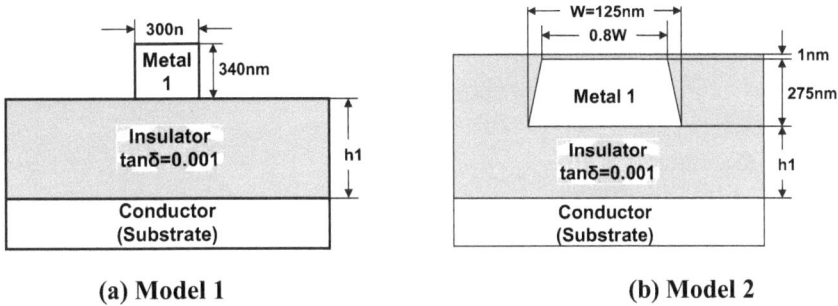

(a) **Model 1**   (b) **Model 2**

Figure 8: Models of On-chip Interconnect.

Condition			Estimation of Characteristic Impedance		
Analyzing Model	h1(m)	Dielectric Constant	MW STUDIO®	APSIM RLGC®	Equation
Model 1	10n	3		204Ω	
Model 2	30n	16	10.14Ω (1GHz) 10.15Ω (20GHz)	190.8Ω (1GHz) 47.0Ω (20GHz)	15.26
		64	5.07Ω (1GHz) 5.07Ω (20GHz)	95.4Ω (1GHz) 23.5Ω (20GHz)	7.72
	0.9μ	16	35.8Ω (1GHz) 38.9Ω (20GHz)	394Ω (1GHz) 95.1Ω (20GHz)	64.4
		32	17.92Ω (1GHz) 17.92Ω (20GHz)	198.0Ω (1GHz) 47.8Ω (20GHz)	32.1
	1.8μ	3.2	-	-	177
1,000 times scaled up of Model 2	30μ	16	10.15Ω (1GHz) 10.15Ω (20GHz)	11.50Ω (1GHz) 11.50Ω (20GHz)	15.26
		64	5.07Ω (1GHz) 5.07Ω (20GHz)	5.75Ω (1GHz) 5.75Ω (20GHz)	7.72
	900μ	16	35.85Ω (1GHz) 35.78Ω (20GHz)	48.7Ω (1GHz) 48.7Ω (20GHz)	62.4
		32	17.89Ω (1GHz) 17.90Ω (20GHz)	24.46Ω (1GHz) 24.46Ω (20GHz)	32.1

Figure 9: Calculation Result of Impedance of On-chip Interconnect.

Fig. **9** shows the calculation result of the impedance of the on-chip interconnect. In Fig. **9**, the commonly-used equation of the microstrip line was considered to be more reliable, because the scaling of the formation does not fit to both simulators. The actual characteristic impedance of the on-chip interconnect including the power supply wiring layers is considered to be being between 210Ω and 177Ω.

From above discussion, the feasibility of the on-chip LILL was confirmed.

DESIGN EXAMPLE OF ON-CHIP LILL

Fig. **10** shows an example of the formation of the on-chip LILL.

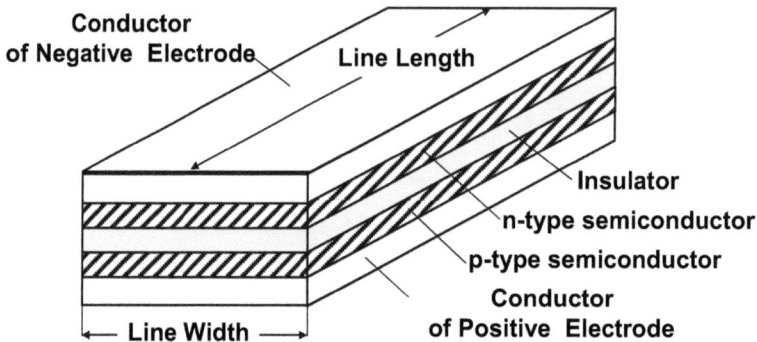

Figure 10: Example of Formation of On-chip LILL.

In Fig. **10**, the attenuation is generated by the n-type semiconductor layer and the p-type semiconductor layer. Each n-type semiconductor layer and p-type semiconductor layer form the Schottky junction with the conductor, and each Schottky junction is biased in reverse. The low impedance is got by the thin insulator layer having the relatively large dielectric constant. The attenuation depends on the line length as well as the conductance of the n-type semiconductor as shown in the developed equations for the on-board LILL. The line length can be shortened because the only pure SEMW exists nearby of on-chip inverter and its wave length is quite short. Therefore forming the on-chip LILL in the on-chip interconnect will become possible.

Fig. **11** shows the calculated characteristics of the on-chip LILL of each line length.

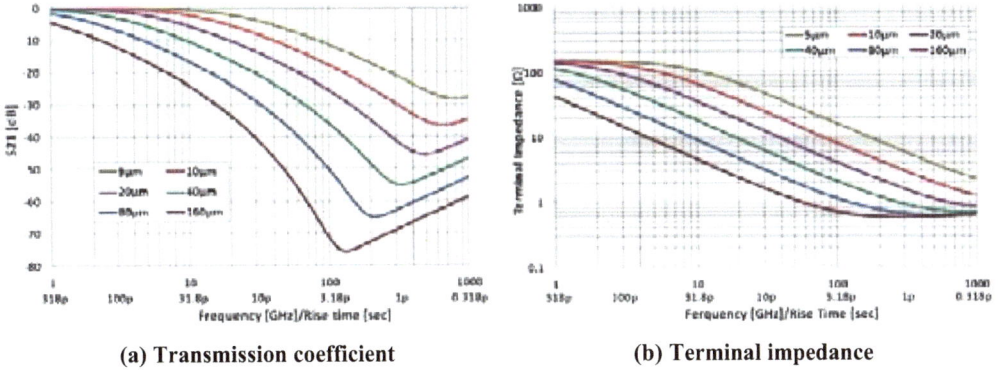

(a) Transmission coefficient   **(b) Terminal impedance**

Figure 11: Calculated Characteristics of On-chip LILL.

In Fig. **11**, the calculation equations which were developed for the on-board LILL were used, the calculation condition was the following; z was to $160\mu m$ from $5\mu m$, σ_S was 10^5 S/m, a was $10nm$, b_S was $100nm$, ε_r was 8.5, μ_r was 1, w was $3\mu m$, C_T was 10^{-22} F/m, and Z_0 was 177Ω, respectively. The rise time was replaced by MSF. The suitable line length will be selected for reducing enough the terminal impedance and the transmission coefficient. The LSI chip is rectangular and $1mm^2$ in area approximately, and the thickness of the on-chip interconnect is larger than $10\mu m$. The on-chip interconnect consists of the metal wire, the insulator, and the ground mesh. Therefore the on-chip LILL can be formed in the interconnection layer.

EFFECTIVENESS

The on-chip LILL is effective for making the rise time of the signal voltage faster because it provides the low-impedance point near the on-chip inverter on the power line. The leakage of SEMW to the signal line on the PCB *via* power line on the chip will be reduced by the on-chip LILL. When the on-chip LILL is used to all LSI and the other semiconductor devices on PCB, the on-board LILL will become unnecessary because the transmission coefficient of the on-chip LILL is as small enough as it of the on-board LILL.

REFERENCES

[1]   H. Tohya, N. Toya, "A Novel Design Methodology of the On-Chip Power Distribution Network Enhancing the Performance and Suppressing EMI of the SoC", IEEE International Symposium on Circuits and Systems 2007 (ISCAS 2007), pp. 889-892, 2007.

[2] H. Tohya, N. Toya, "Design Methodology of On-Chip Power Distribution Network", 2006 IEEE Dallas/CAS Workshop on Design, Applications, Integration and Software, pp. 79-82, 2006.

[3] H. Tohya, N. Toya, "Novel Design Concept and Technology for the Switching Mode Circuit based on the Electromagnetic Wave theory and the Nonlinear Undulation theory", IEEE, ISCIT 2010, pp. 1097-1102.

[4] H. Tohya, N. Toya, "Novel Design Concept and technologies of the Switching Mode Circuit based on the Electromagnetic Wave theory and the Nonlinear Undulation theory". IEEE, TENCON2010, pp. 1135-1140, 2010.

[5] H. Tohya, N. Toya, "Solitary Electromagnetic Waves Generated by the Switching Mode Circuit", http://cdn.intechweb.org/pdfs/15930.pdf

Send Orders for Reprints to reprints@benthamscience.net

CHAPTER 13

Novel MILL Technology

Abstract: The MILL technology became possible to actualize by the SEMW theory. MILL is used to the signal line and it improves the signal integrity and crosstalk greatly.

Keywords: Matched impedance lossy line (MILL), high-speed, reflection, AC circuit theory, SEMW theory, electromagnetic analysis, logic design, lattice diagram, bounce, SMC, signal line, crosstalk, driving terminal, SEW, SEMW, EMI, Fourier Transform, n-type semiconductor, p-type semiconductor, EMI.

ANALYSIS OF REFLECTION ON SIGNAL LINE

The signal bounce which is caused by the reflection is one of the hardest problems for transmitting the signal of the SMC at the high-speed. Conventionally, the reflection phenomenon has been analysed by using the method of the lattice diagram, which has the other names such as the bounce diagram, the echo diagram, or the reflection diagram. This method is being used widely for the design and analysis of the SMC [1, 2]. However this method cannot handle the electromagnetic phenomenon because it is based on the AC circuit theory. The SEMW theory enables the electromagnetic analysis as well as the logic design about the reflection or the bounce on the transmission line of the SMC easily. The technology of MILL was developed to actualize above.

Some analyzed examples of the rising period of the signal voltage by SEMW theory will be shown below. In this analysis, the equivalent circuit shown in Fig. **1** of Chapter **12** is used, VDD which is an ideal voltage source was $3V$, Z1 is the driver, and Z_T is the terminal impedance.

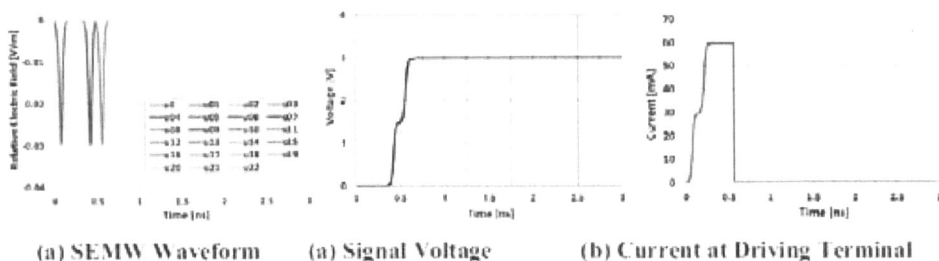

(a) SEMW Waveform   (a) Signal Voltage   (b) Current at Driving Terminal

Figure 1: Waveform and Shepes on Signal Line of First Example.

Fig. **1** shows the waveform and the shapes on the signal line of the first example. Fig. **1a** shows the waveform of SEW. Fig. **1b** shows the signal voltage shape. Fig. **1c** shows the signal current shape at the driving terminal C.

Fig. **2** shows the waveform and the shapes on the signal line of the second example. Fig. **2a** shows the waveform of SEW. Fig. **2b** shows the signal voltage shape. Fig. **2c** shows the signal current shape at the driving terminal C. The calculation condition was the following; each length of the power line and the signal line was 10*mm* and 20*mm*, Z_T was 250Ω at MSF of the switching time of 0.1*ns*, and others were same as the first example.

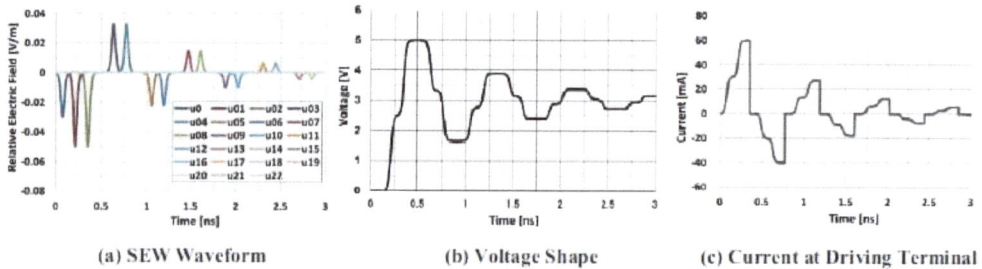

(a) SEW Waveform   (b) Voltage Shape   (c) Current at Driving Terminal

Figure 2: Waveform and Shepes on Signal Line of Second Example.

In Fig. **2a**, the waveforms from u01 to u22 are SEW which are observed at the terminal of the signal line, each odd number of the waveform corresponds to SEW generated on the signal line, each even number of the waveform corresponds to the reflection wave of the SEW generated on the power line, the amplitude of u01 and u02 is twofold of u0, and the magnitude of other SEW is relatively large. These SEWs having large amplitude tend to cause the crosstalk and EMI against the other circuit. In Fig. **2b**, the voltage wave shape is got by the integral of the wave shape of the SEWs shown in Fig. **2a**, the rise time caused by the first bounce was 0.164*ns* which is 1.64 times of the switching time of the driver. However, the actual rise time increases to 0.92*ns* which is 92 times of the switching time of the driver because the signal voltage of the second bounce is near to the threshold voltage of the receiver. In Fig. **2c**, the necessary time for charging the signal line is 0.35*ns*, and the current pulses for the charge and the discharge exist till the bounce disappears.

Fig. **3** shows the waveform and the shapes on the signal line of the third example. Fig. **3a** shows the waveform of SEW, Fig. **3b** shows the voltage shape, and Fig. shows the current shape at the driving terminal C. The calculation condition was the following; ZT was 20Ω at MSF which corresponds to the switching time of $0.1ns$, and others were same as the second example.

In Fig. **3a**, the waveforms from u01 to u22 are SEW which are observed at the terminal of the signal line.

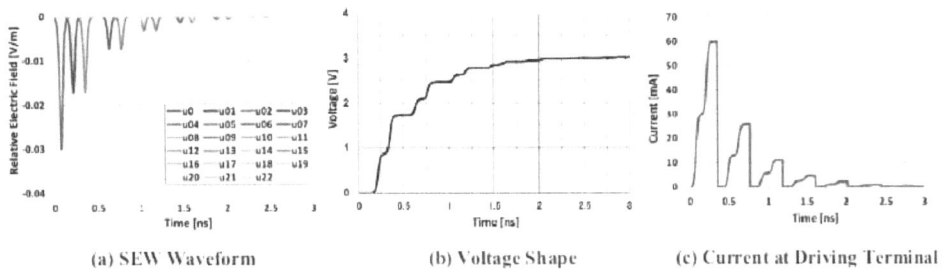

(a) SEW Waveform   (b) Voltage Shape   (c) Current at Driving Terminal

Figure 3: Waveform and Shepes on Signal Line of Third Example.

In Fig. **3**, each odd number of the waveform corresponds to SEW generated on the signal line. Each even number of the waveform corresponds to the reflection wave of the SEW generated on the power line. The amplitude of u01 and u02 is 0.57 times of u0. The other SEMW exist but the magnitude of them is relatively small. These SEW has the possibility to cause the crosstalk and EMI against the other circuit. In Fig. **2b**, the voltage wave shape is got by the integral of the wave shape of the SEWs shown in Fig. **3a**, the rise time is $0.98ns$ which is 98 times of the switching time of the driver. In Fig. **3c**, the necessary time for charging the signal line is $2.5ns$ and no current pulses exist after the finishing time of charging.

PERFORMANCE OF MILL ON SIGNAL LINE [3, 4]

EMI and the crosstalk do not occur on the DC circuit or the low-frequency circuit. According to the electromagnetism, the AC circuit including the SMC is the EMW circuit. EMI and the crosstalk on the signal line will be reduced, if EMW on the signal line is reduced nearby of the driving terminal. The SEMW theory presents the technology for actualizing it as MILL.

Fig. **4** shows the equivalent circuit of SMC including the LILL and the MILL. MILL is the abbreviation of the matched impedance lossy line, which is formed by the transmission line structure. MILL has the same characteristic impedance to the connected signal line and it has the relatively large attenuation. MILL is connected between the point D and E.

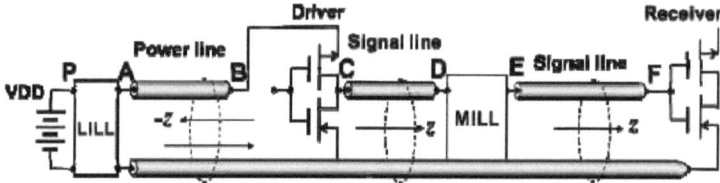

Figure 4: Equivalent Circuit of SMC Including LILL and MILL.

In Fig. **4**, the length of the signal line between the point C and D should be short but the length of the signal line between the point E and F is not limited. LILL is connected between the point P and A. The length of the power line between the point P and A is not limited. The resistance of MILL is negligible. Each SEMW is generated on the power line and the signal line at the switching period of the driver. The signal voltage is formed by both SEW generated on the signal line and SEW generated on the power line.

The magnitude of SEW is reduced exponentially on the MILL and no reflections are occurred at the point D and E. The magnitude of the signal voltage is not reduced in spite of such behavior of SEW. If the magnitude of the signal voltage is reduced, the DC voltage drop between point D and E becomes large. However, the resistance of MILL is negligible. Therefore such phenomenon never occurs, because the driver cannot feed the so large DC current to MILL.

Fig. **5** shows the behavior of SEW on the MILL.

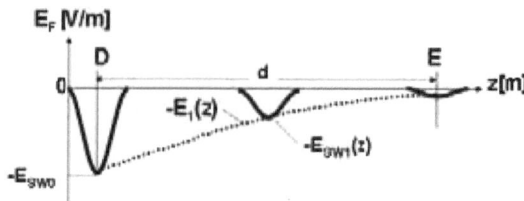

Figure 5: SEW on the MILL.

In Fig. **5**, the magnitude of the charge voltage is reduced according to the reduction of the magnitude of SEW on MILL. However, the exponential variation which is shown by the dash line fully compensates for the reduction of the charge voltage on MILL.

The electric field strength of the SEW on the MILL is:

$$E_{SW1}(y,z) = e^{-\alpha z}E_0 E_{SW}(y,z) \qquad (1)$$

where each y is the axis if the thickness and z is the axis the length, α is the attenuation constant. $E_0 E_{SW}(x,z)$ is the input electric field of the SEW.

According to the definition of the electromagnetism, the signal voltage on the MILL is:

$$V_1 = -\int\int -E_{SW1}(x,z)\partial y \partial z \qquad (2)$$

In the equation 2, when x is d which is the effective thickness of the insulator of MILL and z is l which is the line length of MILL

$$V_1 = -V_s(-e^{-\alpha l}) = V_s e^{-\alpha l} \qquad (3)$$

The integral of the variation form of the magnitude of SEW which is shown the dash line in Fig. **5** is:

$$V_2 = V_s(1 - e^{-\alpha l}) \qquad (4)$$

From the equation 3 and 4, the sum total of the signal voltage is:

$$V_1 + V_2 = V_s \qquad (5)$$

From the equation 5, it is understood that the signal voltage keeps the constant value V_S on the MILL.

As the result, the crosstalk and the bounce on the signal line between MILL and the receiver will be reduced, and the signal integrity will be kept because SEMW is attenuated enough by MILL.

According to the conventional circuit theory depending on the Fourier Transform, the loss of the signal line should be minimized because of degrading of the signal integrity. However, this idea was corrected by the SEMW theory, and using the transmission line having the large attenuation to the signal line of SMC became possible to actualize for the first time in the world.

DESIGN EXAMPLES OF MILL [3, 4]

Fig. **6** shows the formation of the first design example of MILL.

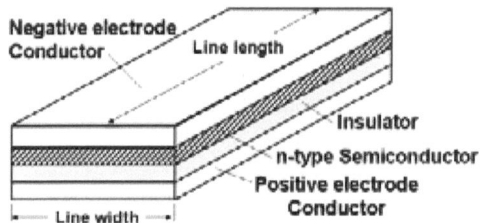

Figure 6: Formation of First Design Example of MILL.

In Fig. **6**, MILL is formed by the parallel plate transmission line by placing the n-type semiconductor layer between the insulator and the conductor. Fig. **7** shows the calculated characteristics of the first design example of MILL. Each line length of MILL01, MILL02, MILL04, MILL08, MILL12, MILL18, MILL27, and MILL41, is 1*mm*, 2*mm*, 4*mm*, 8*mm*, 12*mm*, 18*mm*, 27*mm*, and 41*mm*, the transmission coefficient of -20dB is considered to be effective, and the reflected coefficient of all sample is considered to be enough. The characteristic impedance of these samples was designed to 50Ω. The calculation condition was the following; the n-type semiconductor was 700S/m, each thickness of the insulator and the n-type semiconductor was 13μm and 30μm, ε_r of the insulator was 3.5, line width was 55μm, and the coupling capacitance between the terminals was 10^{-18} F/m, respectively.

Fig. **8** shows the formation of the second design example of MILL. The second example of MILL consists of the n-type semiconductor layer and the p-type semiconductor layer, and the formation of the MILL is similar to it of the on-chip LILL, and the characteristic impedance MILL is designed to match to it of the signal line.

(a) Transmission coefficient

(b) Reflection Coefficient

(c) Terminal Impedance

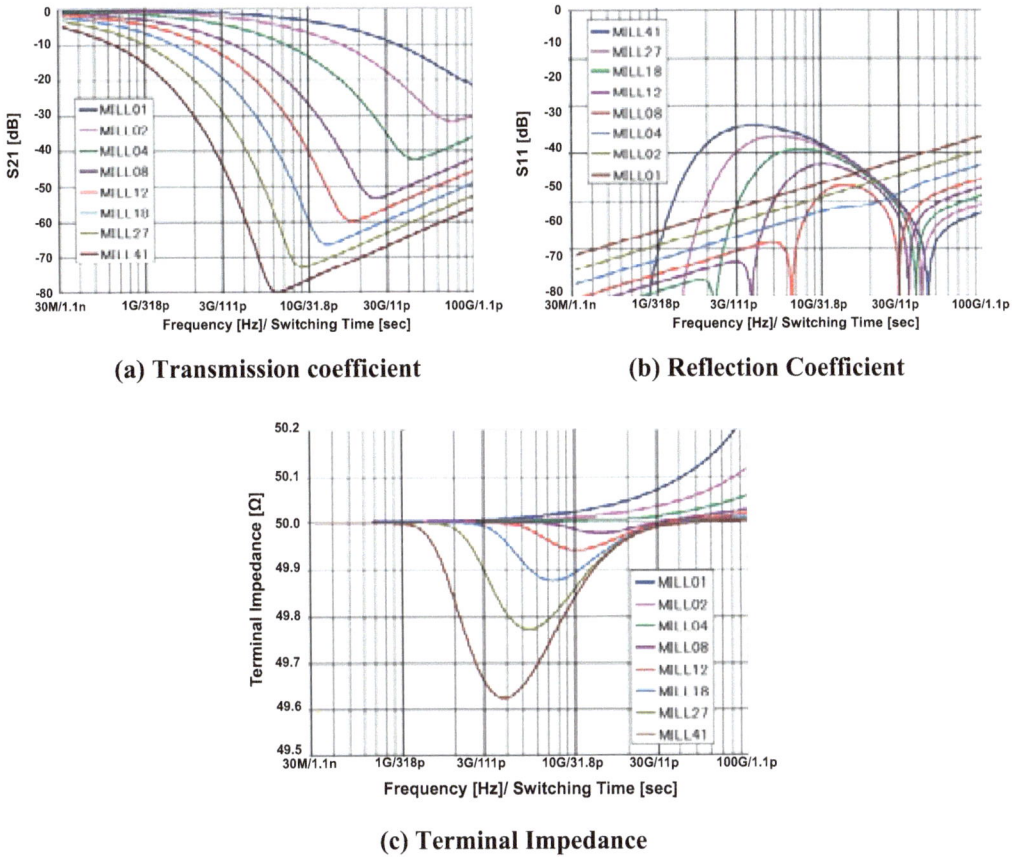

Figure 7: Calculated characteristics of the design example of the MILL.

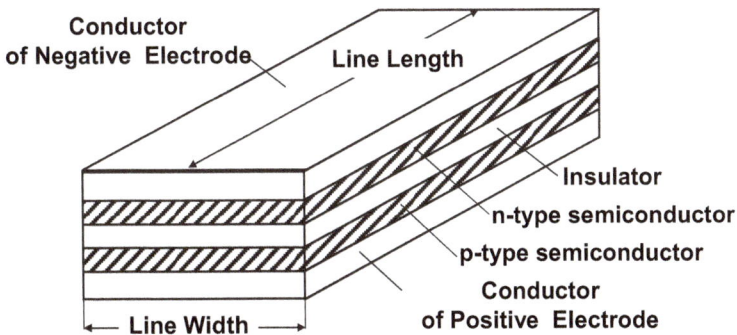

Figure 8: Formation of Second Example of MILL.

Fig. **9** shows the calculated characteristics of the second design example of the MILL.

(a) Transmission Coefficient   **(b) Reflection Coefficient**

(c) Terminal Impedance

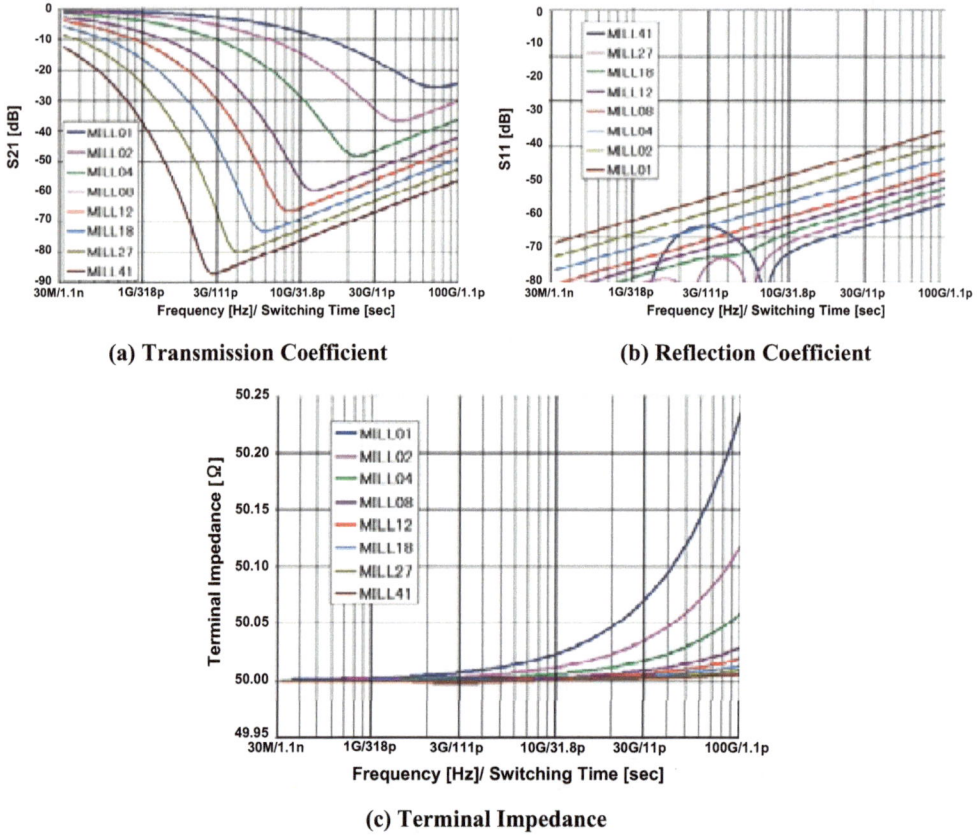

Figure 9: Calculated Characteristics of Design Example of MILL.

In Fig. **9**, the transmission coefficient and the reflected coefficient of the first example are improved. The calculation condition was the following; the conductivity of both n-type and p-type semiconductor were $4,000 S/m$, each thickness of the insulator layer and the semiconductor layer was $5\mu m$ and $30\mu m$, ε_r of the insulator was 3.5, line width was $20\mu m$, and the coupling capacitance between the terminals was 10^{-18} F/m, respectively. In Fig. **9a**, the transmission coefficient is considered to be about 20dB, because the crosstalk and the bounce at the terminal without the impedance matching circuit are reduced enough, and the EMI by the signal line is quite fewer than EMI by the power line.

The line width of the designed MILL is small enough. Therefore, MILL component will be formed which consists of many integrated chips and the shielding barriers between chips.

VALIDATION OF PERFORMANCE [3, 4]

Prototyping of Lab Sample of MILL

Fig. **10** shows the prototyped lab sample of MILL for the validation of above mentioned performance of MILL.

Figure 10: Prototyped Lab Sample of MILL.

In Fig. **10**, the lab sample of MILL was made with the carbon paste, with the silver paste, and with the polyurethane enamelled wire of $0.1mm\Phi$ which has the insulator of $5\mu m$ thickness. The thickness of the carbon graphite layer was $0.1mm$ approximately. The line length was $100mm$ and the characteristic impedance was designed to 6.8Ω because prototyping of the first and second design samples were difficult. The thickness of the insulator of the polyurethane enamelled wire was the minimum value in the commercialized insulated wire.

Fig. **11** shows the measured characteristics of the lab sample of MILL for the experiment.

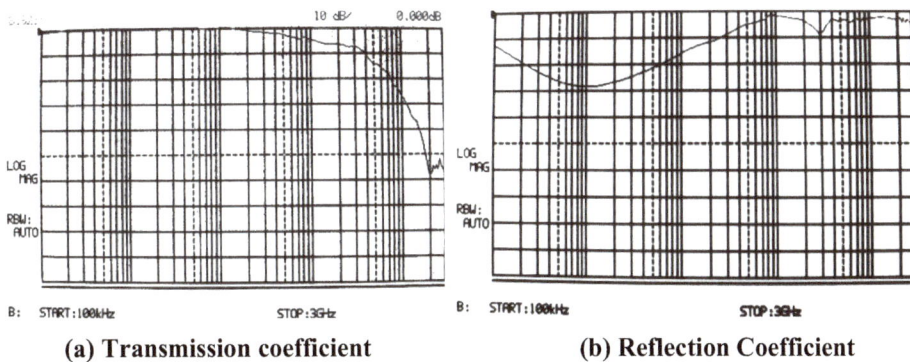

(a) Transmission coefficient   **(b) Reflection Coefficient**

Figure 11: Measured Characteristics of Lab Sample of MILL.

In Fig. **11**, the *x*-axis shows the frequency allocated to $3GHz$ from $100\ kHz$ on the log scale. The *y*-axis is allocated to -100dB from 0dB on the linear scale. The characteristics of the lab sample were measured by the network analyzer. The measured characteristic of the lab sample of MILL was the limit of the hand made by using the current materials.

Fig. **12** shows the calculated characteristics of the lab sample of MILL.

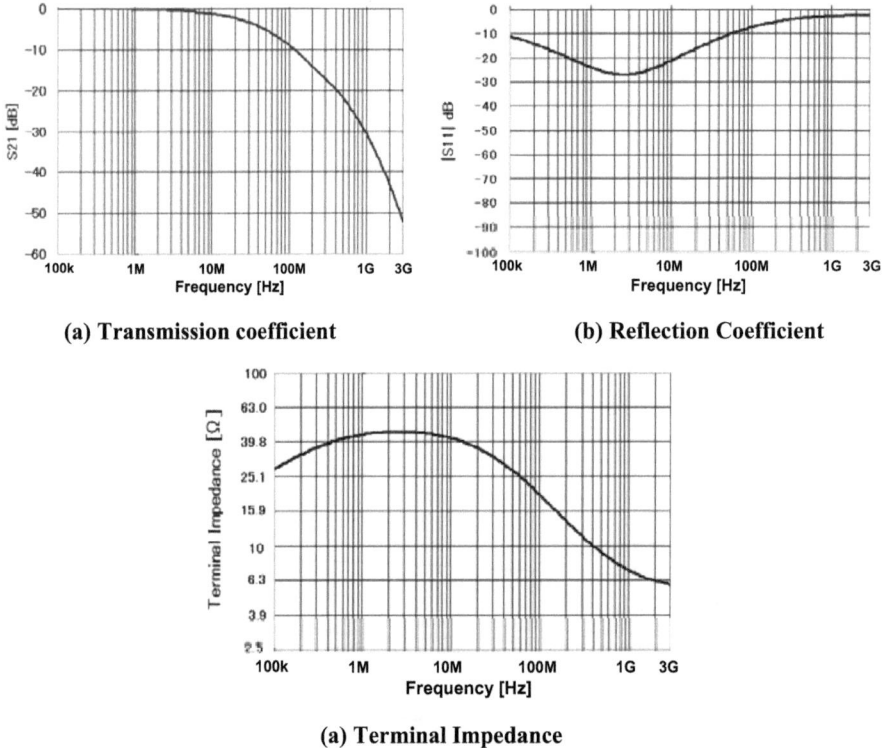

(a) **Transmission coefficient**   (b) **Reflection Coefficient**

(a) **Terminal Impedance**

Figure 12: Calculated Characteristics of Lab Sample of MILL.

In Fig. **12**, the calculation condition was the following; the conductivity of both n-type and p-type semiconductor were $4,000 S/m$, each thickness of the insulator layer and the semiconductor layer was $5\mu m$ and $30\mu m$, ε_r of the insulator was 3.5, line width was $20\mu m$, and the coupling capacitance between the terminals was 10^{-18} F/m, respectively.

Validation of Effectiveness by Experiment

Fig. **13** shows the prototyped test board. The oscillators and the gate drivers are same as these which were shown in Fig. **6** in Chapter **12**. Four $14mm$ prototypes of the on-board LILL and one lab sample of MILL were used. The transmission lines were formed by two couples of the polyethylene wires which was placed directly on the ground plane. And the characteristic impedance of the transmission lines was designed to 50Ω.

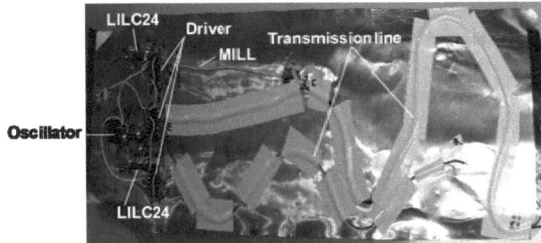

Figure 13: Prototyped Test Board.

Fig. **14** shows the circuit diagram of the prototyped test board. The impedance matching resistors are not use to the terminal of the signal line.

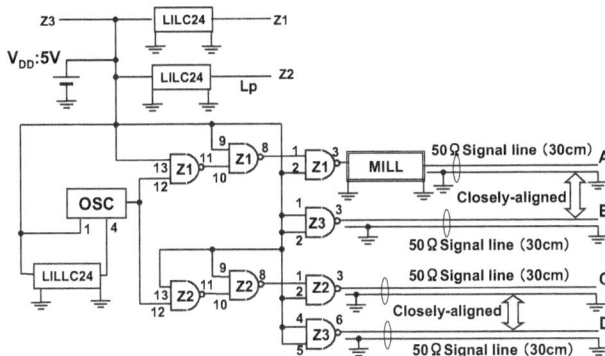

Figure 14: Circuit Diagram of Prototyped Test Board.

Fig. **15** shows the measured voltage wave forms of the point from A to D shown in Fig. **14**. Fig. **15a** shows the signal voltage of the point A. Fig. **15b** shows the crosstalk voltage at the point B. Fig. **15c** shows the signal voltage at the point C. Fig. **15d** shows the crosstalk voltage at the point D.

In Fig. **15**, the bounce on the signal voltage is little observed when MILL is used but the large bounce is observed when MILL is not used. The magnitude of the signal voltage when MILL is used is larger than the case of non-use of MILL. The crosstalk voltage becomes negligible in actual use when MILL is used but the crosstalk voltage is too large to transmit the signal when MILL is not used.

(a) Voltage Shape at Point A

(b) Voltage Shape at Point B

(c) Voltage Shape at Point C

(d) Voltage Shape at Point D

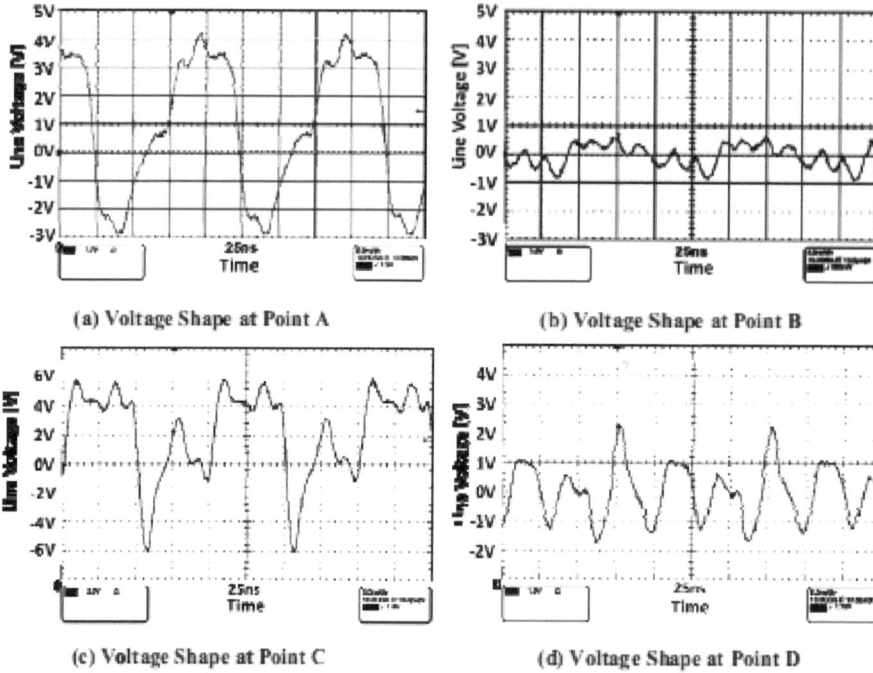

Figure 15: Measured Voltage Shape of Signal Line.

CONCLUSIONS

The damping resistor is used usually for reducing the bounce. However it cannot reduce the cross talk because the series connected resistor cannot attenuate the traveling EMW in the insulator of the transmission line effectively. The appealing performance of the MILL was presented. The signal integrity is maintained and the crosstalk and the bounce are suppressed greatly even if the impedance matching circuit is not used to the receiving terminal of the signal line when the MILL is used on the signal line. In addition, EMI caused by the signal line is considered to be suppressed enough.

REFERENCES

[1]   C.W. Trueman, "Animating transmission-line transients with bounce", IEEE, Transactions on Education, pp. 115–123, Vol. 46, Issue 1, 2003.

[2]   C.-T. Tsai, "Signal integrity analysis of high-speed, high-pin-count digital packages", IEEE, Electronic Components and Technology Conference, vol. 2, pp. 1098–1107, 1990.

[3]   H. Tohya, N. Toya, "Novel Design Concept and technologies of the Switching Mode Circuit based on the Electromagnetic Wave theory and the Nonlinear Undulation theory". IEEE, TENCON2010, pp.1135-1140, 2010.

[4]   H. Tohya, N. Toya, "Solitary Electromagnetic Waves Generated by the Switching Mode Circuit", http://cdn.intechweb.org/pdfs/15930.pdf

Send Orders for Reprints to reprints@benthamscience.net

CHAPTER 14

Innovative Circuit and System Technologies for SMC

Abstract: The reconfiguration of SMC to QSCC will actualize when the LILL technology, MILL technology, and the SEMW theory is applied.

Keywords: QSCC, red brick wall, trapezoidal voltage shape, Fourier Transform, AC circuit theory, telegrapher's equations, charge current, timing skew, circuit simulator, CAD, size, price, turn-around time, signal integrity, wiring criteria, parallel-bit high-speed transmission, SMPS circuit, buck converter, spike, SerDes.

PROBLEMS OF CONVENTIONAL THEORY AND TECHNOLOGIES

It is considered that the first SMC was used to the telecommunication equipment using the Morse code. The first general-purpose digital computer as ENIAC consists of SMC. After that, SMC has been improved greatly by the support of the improvement of the performance of the semiconductor technologies. Almost all equipment and apparatus as well as the society in the world have been innovated by IT which is supported by SMC. However IT is encountering the temporary lull of the growth today. Many causes are thought. One of the dominant causes was considered to be the red brick wall which consists of the conventional theories and the technologies.

The first problem is the conventional theories and the technologies which use the continuous waves. The original of the continuous waves such as the voltage wave and the current wave is EMW. According to the electromagnetism, EMW is generated when the electric field or magnetic wave is changed. These are changed at only switching period of SMC. Therefore, the switching device does not generate the continuous wave but the intermittent wave. The idea that the switching voltage shape such as the trapezoidal voltage shape on SMC consists of many harmonics is not corresponding to the physics. The switching voltage shape has two stationary states usually. However, one stationary state is got by the Fourier Transform, which depends on the variable duty cycle. It can be said that the idea of Fourier transform does not correspond to the physics also from this fact.

Hirokazu Tohya

The secondary problem is that the AC circuit theory and the telegrapher's equations depend on the charge current which drifts at the speed of the order of a millimeter per second. The drift speed of the charge current is not serious at the design of the conventional AC circuit because it is performed by the continuous low-frequency EMW. However, the drift speed or the travel speed is serious at the design of the digital circuit because the management of the timing skew is one of the important items for the high-performance digital circuit. The telegrapher's equations are also depending on the charge current, which were developed by Oliver Heaviside in 1880s for the telecommunication engineering. He recast the modernized versions of the Maxwell's equations which are formed by four equations from twenty equations in 1884. Therefore, he would definitely understand the differences between the electric current in accordance with the Ampère's circuit law and the charge current. However the telecommunication engineers depending on the telegrapher's equations have designed by using the charge current. The skin effect which is not being permitted in the electromagnetism is presented in the telecommunication engineering about the charge current. The Litz wire which is used to the high-frequency power circuit is based on this idea. Conventionally, the microwave circuit as well as the high-speed SMC has been following the telecommunication engineering.

BREAKING RED BRICK WALL OF CONVENTIONAL TECHNOLOGIES

The advocating SEMW theory will be able to break the red brick wall perfectly. The developed technologies of LILL and MILL based on the SEMW theory can reconfigure the conventional SMC to QSCC. SEMW theory and the technologies of LILL and MILL will improve SMC greatly. The example of the improvement items of the equipment and the apparatus was the following;

(Design and Analysis)

To suppress EMI, the cross talk, and the bounce.

To improve the quality by applying the SEMW theory to the circuit simulator, the electromagnetic simulator, and the CAD tool.

To make easy the design of the high-performance SMC.

To increase the flexibility of the figuration of the equipment and the apparatus.

(Products)

To increase the performance and the function.

To reduce the size and weight.

To reduce the price.

(Manufacturing)

To reduce the cost by eliminating almost all electric components and the shielding material.

To reduce the turn-around time (TAT).

The examples of the reconfiguration method of SMC to QSCC will be presented in this chapter.

BAKOGLU'S IDEA

The displacement current or EMW is considered to be dominant media on the transmission line which should be handled as the distributed element model. On the other hand, the considering of displacement current or EMW is not necessary when the length of the transmission line is short enough. The limit of the transmission line on SMC should be decided in accordance with the SEMW theory. The alternative convenient judgment criteria of it will be the Bakoglu's idea [1]. The summary of is the following;

The first condition: $t_r < 2.5t_f$; the transmission line should be handled as the distributed element model in this case.

The second condition: $t_r > 5t_f$; the transmission line can be handled as the lumped element model in this case.

The third condition: $2.5t_f < t_r > 5t_f$; the transmission line can be handled as either the distributed element model or the lumped element model in this case.

where t_r means the rise time which is defined as the time required for the signal to move from 10 percent to 90 percent of its final value, and t_f means traveling time on the transmission line.

Above mentioned Bakoglu's idea was validated by ApsimRLGC which is the commercialized simulator of the transmission line.

Fig. **1** shows the model and the equivalent circuit of the on-chip interconnect. Fig. **1a** shows the model of the transmission line for. Fig. **1b** shows the equivalent circuit of the transmission line model for the simulation by ApsimSPICE. The dielectric constant was 3. Each inductance and capacitance was set to $0.958nH/mm$ and $25.01fF/mm$.

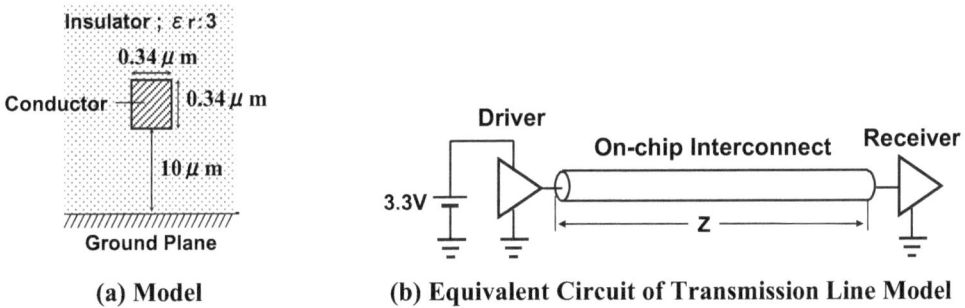

| (a) Model | (b) Equivalent Circuit of Transmission Line Model |

Figure 1: Model and Equivalent Circuit of On-chip Interconnect.

The characteristic impedance was set to 204Ω. And the calculated traveling time was $5.8ps/mm$. In Fig. **1b**, the on-chip interconnect was given as the sub-circuit, the rise time of the driver was $10ps$, the output resistance of the driver was set to 0Ω, the source voltage was set to $3.3V$, the gate capacitance of the receiver was set to $0.03pF$, and the length z of the on-chip interconnect was set to $0.5mm$, $1mm$, and $5mm$.

Fig. **2** shows each current and voltage shapes of the rising part at the driver when the line length is $0.5mm$. Fig. **2a** shows the driver current. Fig. **2b** shows the receiver voltage. The current and the voltage are not bouncing. The traveling time is $2.9ps$ when the line length is $0.5mm$, and it corresponds to the second situation of the Bakoglu's idea. Therefore, this transmission line can be handled as the lumped element model and it can form QSCC barely.

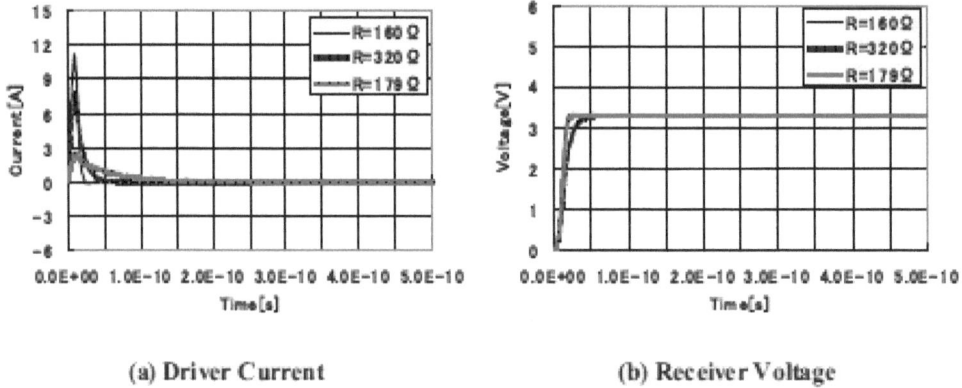

(a) Driver Current   (b) Receiver Voltage

Figure 2: Each Current and Voltage Shapes of Rising Part at Driver when Line Length is 0.5*mm.*

Fig. 3 shows each current and voltage shapes of the rising part at the driver when the line length is 1mm. Fig. 3a is the driver current. Fig. 3b is the receiver voltage.

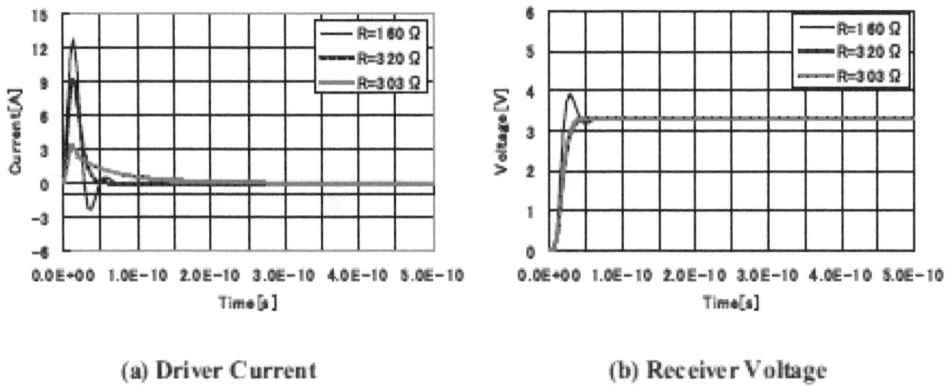

(a) Driver Current   (b) Receiver Voltage

Figure 3: Each Current and Voltage Shapes of Rising Part at Driver when Line Length is 1*mm.*

In Fig. **3**, the current and the voltage are bouncing slightly. The traveling time is 5.8*ps* when the line length is 1*mm*, and it corresponds to the first situation of the Bakoglu's idea. Therefore, this transmission line should be handled as the distributed element model and it cannot form QSCC. From above discussion, the Bakoglu's idea was confirmed to be able to be applied to reconfigure SMC on the chip and it on PCB to QSCC.

NOVEL WIRING CRITERIA FOR FORMING QSCC

QSCC of SMC will be formed by applying the LILL and MILL technologies to the transmission line and the wire in accordance with the following criteria.

(Criteria for Forming QSCC)

1st : The Bakoglu's idea is adopted for the judgment. The rise time (t_r) corresponds to the effective interval (t_s) of SEMW.

2nd : QSCC is considered to be formed when SMC is designed based on the second condition shown in the Bakoglu's idea.

3rd : The limit length of the transmission line for forming QSCC can be relaxed by the magnitude of SEMW in each individual case when the third condition shown in the Bakoglu's idea exists in the partial part of SMC.

4th : When MILL fills the second condition shown in the Bakoglu's idea and it has the enough attenuation performance, the signal integrity can be analyzed by handling MILL as the lumped element model.

RECONFIGURATION EXAMPLES OF SMC TO QSCC

On-Chip Interconnect

On-chip LILL is effective for forming PDN on the chip to QSCC and it is effective for shorten the rise time of the signal voltage.

The characteristic impedance which was considered to be 200Ω approximately is effective for keeping the small power consumption of LSI. However, it makes forming the on-chip MILL difficult because the large attenuation due to the absorption loss cannot be gotten on the high-impedance transmission line. Therefore, the repeater is considered to be effective for the signal line on the chip. By using the repeater, QSCC of the signal line will be able to be formed by limiting the length of the signal line in accordance with the criteria for forming QSCC. The limit length of the on-chip interconnect is each $42\mu m$ and $12.7\mu m$

when the switching time of the driver is $1ps$ and $0.3ps$. The characteristic impedance of the signal line and the power line can be increase to near of $120\pi\Omega$ which is the intrinsic impedance of vacuum when the on-chip interconnect is reformed to QSCC. The power consumption of LSI will be more reduced by this.

From above, QSCC of LSI will be successfully formed.

PCB of Digital Equipment

The on-board LILL is effective to forming QSCC when it is used to PDN on PCB. The on-board LILL should be used to each nearby power terminals of LSI and power source. The on-board MILL is effective to forming QSCC when it is used to the signal line on PCB. The on-board MILL should be used to each nearby terminal of the signal driver and the signal receiving connector on PCB.

Fig. **4** shows an example of the cross section of PCB of digital equipment for reconfiguring PDN to QSCC. LILL is used at the nearby place of LSI. The traces between the power terminals of LSI and LILL are formed as short length as possible. The vias between the power terminals of LILL and LSI are made to resemble the transmission line. And the bulk capacitors are used for the electrostatic stability. LILL should be also placed at the nearby place of the power receiving terminal.

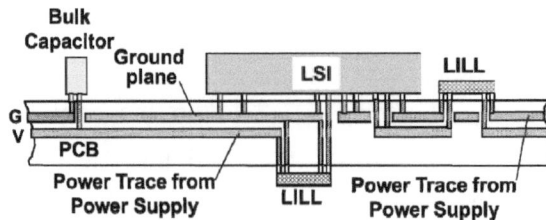

Figure 4: Example of Cross Section of PCB for Reconfiguring PDN to QSCC

The characteristic impedance of the power lines between LILL and the power supply should be relatively large to get the large attenuation of LILL. The limit length of the trace between the power terminals of LSI and LILL for forming QSCC is $9.2mm$ at $1GHz$ and $1.5mm$ at $6GHz$. As the result, PDN will be successfully reconfigured to QSCC.

Fig. **5** shows an example of the cross section of PCB for reconfiguring the signal lines to QSCC. MILL is used at the nearby place of LSI.

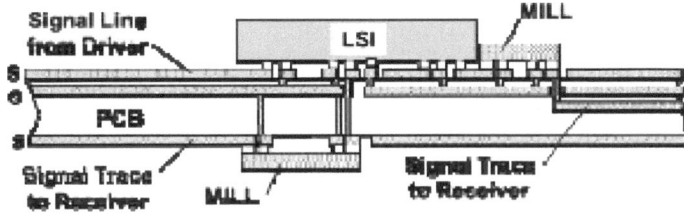

Figure 5: Example of Cross Section of PCB for Configuring Signal Lines to QSCC.

The traces between the driver terminals of LSI and signal terminals of MILL are formed as short length as possible. The vias between the driver terminals of LSI and the signal terminals of MILL are made to resemble the transmission line. The characteristic impedance of the signal lines between MILL and the receiving terminals is desirable to be relatively large to get the large attenuation of MILL and to reduce the power consumption of the driver. The limit length of the trace between the driver terminals of LSI and the signal terminals of MILL for forming QSCC is each 2.9*mm* and 0.86*mm* when the switching time of the driver is 100*ps* and 30*ps*. As the result, signal lines will be successfully formed to QSCC. MILL should be placed at the nearby place of the LSI.

The driver in LSI generates SEMW when it changes ON/OFF. The negative polar SEMW is generated on the signal line when the driver changes ON. The positive polar SEMW generated on the power line at the same time is reflected at LILL and it changes its polarity. The reflected negative polar SEMW follows the SEMW which was generated on the signal line. The signal line is charged by two negative polar SEMW which pull out the electrostatic energy from the DC power supply. The signal line is discharged to the zero by the positive polar SEMW which is generated by changing OFF of the driver. The electrostatic energy charged is consumed by the on-chip inverter of LSI at this time.

In Fig. **5**, the large numbers of transistor in LSI generate SEMW by their changing ON/OFF. LILL used at the nearby place of LSI reflects almost of SEMW toward LSI. The reflected SEMW forms a part of the signal. Therefore, SEMW little exist on the power line to the power supply from LILL. EMI on the power line including the vias will become negligible when the wire length is shortened enough in accordance with the criteria for forming QSCC. The signal integrity of LSI will be kept because the terminal impedance of the LILL is small enough.

The signal line consists of the excellent transmission line such as the microstrip line and they are formed independently. Therefore, the contribution ratio of SEMW on the signal lines is smaller than it on the power line. The magnitude of SEMW on the signal line is decreased by MILL effectively when it is used in a way of Fig. **5**. EMI about the signal line including the crosstalk and the bounce will become negligible by this.

From above, QSCC of the whole of the digital circuit on PCB will be successfully formed. As the result, the design of the digital circuit having the high-performance and many functions will become easy.

The dumping resistor is used commonly for reducing the bounce, however its effect is limited and it cannot reduce the crosstalk on the high-speed transmission line. The impedance matching technique for the high-performance transmission line becomes unnecessary when MILL used. The serial-bit transmission such as the SATA and USB has been used currently because of the crosstalk. However, SEMW theory and the technologies of LILL and MILL enable the parallel-bit high-speed transmission. The long distance high-speed data transmission will be actualized easily by this.

SMPS Circuit

The SMPS circuit is a kind of SMC and it is designed and analysed by the AC circuit theory except the operation analysis. Therefore, EMI problem cannot solve effectively. The output of the SMPS circuit is the regulated DC voltage or current or regulated AC voltage. The state space averaging method (SSA method) is applied to the stationary state analysis of the SMPS circuit. The switching ripple is smoothed by the inductor and the capacitor. The current design method of the regulation control circuit becomes possible by applying the SSA method. In the SSA method, the state-equations consist of the average voltage or current which are proportional to the duty cycle of the switching voltage. The topological analysis of SMPS becomes easy by applying the SSA method. The regulation of the output of SMPS is controlled by negative feedback usually. The response time of the regulation control is set to several hundred times of the switching frequency usually for keeping its stability. The noise filter is used to the receiving terminal for the defence against the electromagnetic conductive disturbance on the power transmission line.

The key factors of improvement are the following; improvement of the conversion efficiency, suppression of EMI, improvement of the responsibility of the load change and the source voltage change, and reduction in size and weight. The improvement of the switching transistor tends to increase the working voltage and current and to reducing the conversion loss instead of increasing the switching frequency. For these objectives, Si MOSFET and IGBT (insulated gate bipolar transistor) have been improved, and SiC MOSFET and GaN transistor were developed and were being improved. The conducted emission which is caused by the switching transistor is limited at the mains port by IEC/CISPR. The harmonic waves on the mains are generated by rectifier circuit of the front end and these are limited by EMC standard of IEC. Therefore, the switching frequency of SMPC which is connected to the mains has been set to be lower than 150 kHz, and the power factor correction circuit which consists of the boost converter is used in between the rectifying circuit and the converter for reducing the harmonic waves.

LILL and MILL are also Effective for the SMPS Circuit

Fig. **6** shows an example of the reconfiguration of the back converter which is reconfigured to QSCC by using LILL and MILL.

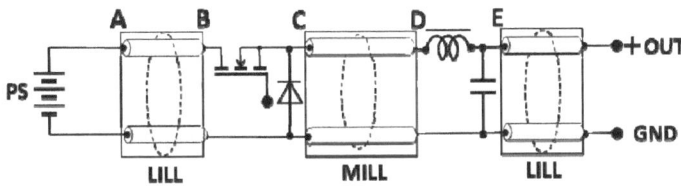

Figure 6: Example of Back Converter Reconfigured to QSCC by LILL and MILL.

In Fig. **6**, PS is the DC voltage source, LILL is connected between the point A and the point B, NMOSFET is connected between the point B and C, the flywheel diode is connected to the point C and the ground, MILL is connected between the point C and D, the inductor is connected between the point D and E, the capacitor is connected between the point E and the ground, and the other LILL is connected between the point E and the output terminal.

NMOSFET generates SEMW when it changes ON/OFF. The negative polar SEMW is generated on the source of NMOSFET when NMOSFET changes ON. The positive polar SEMW generated on the drain of NMOSFET at the same time is reflected at LILL and is changed its polarity. The reflected negative polar SEMW follows the SEMW which was generated on the source of NMOSFET. The transmission line from source of NMOSFET to the point C is charged by two negative polar SEMW which pull out the electrostatic energy from the DC power supply.

SEMW is attenuated in MILL. However, about the voltage waveform, the amplitude and the switching time are maintained in MILL. The magnitude of SEMW of between the point D and E becomes relatively small. SEMW is not necessary for the smoothing circuit because it consists of the passive component. The magnitude of SEMW at the output terminal will become negligible when LILL is used to the output circuit. The positive polar SEMW is generated at the period of changing OFF of NMOSFET because the flow of the electrostatic energy is shut down. The transmission lines between the source of NMOSFET and the output are discharged by the generated positive polar SEMW. The electrostatic energy in the inductor and MILL is discharged by the diode at the period of changing OFF of NMOSFET. Unlike the digital circuit, the consumption of the electrostatic energy at changing OFF of NMOSFET is little.

EMI on the transmission line of the back converter except LILL and MILL becomes negligible when they are formed by the short wire in accordance with the criteria for forming QSCC. When RJK0368DPA (Renesas) is used as NMOSFET, the typical rise time is $3.5ns$ and the internal impedance of NMOSFET is 1Ω. In Fig. **6**, the lumped element model except the attenuation characteristic of MILL can be adopted when the effective length of the chip of MILL is filled the second criteria for forming QSCC. The maximum effective length is $12mm$ when MILL is made of etched aluminum film at this application.

Fig. **7** shows the characteristics of the design example of LILL for SMPS.

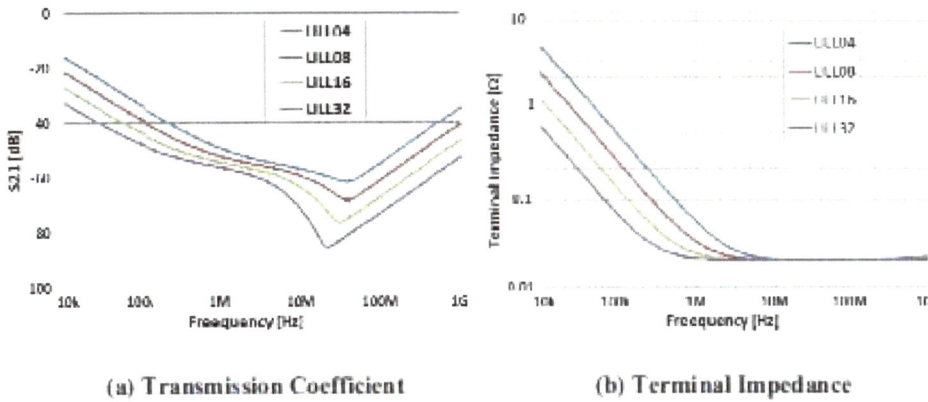

(a) Transmission Coefficient   (b) Terminal Impedance

Figure 7: Characteristics of Design Example of LILL for SMPS.

In Fig. 7, the calculation condition was the following; z was to $32mm$ from $4mm$, C_1 was $95.4\mu F$, R was 0.9, k was 224.7, L was 0.3, w was $2mm$, a was $35.4nm$, ρ_C was $300\mu\Omega m$, each σ of the conductive polymer and the carbon graphite was $12,000S/m$ and $72,727S/m$, ε_r of the alumina was 8.5, b_A for the calculation of the characteristic impedance of the attenuation constant was 0, b_A for the calculation of the characteristic impedance of the reflection coefficient was $0.2\mu m$, b_s was $1\mu m$, b_c was $2.29\mu m$, and C_T was $1.2\times10^{-16}F$. The thickness of the etched aluminum film was $110\mu m$, respectively.

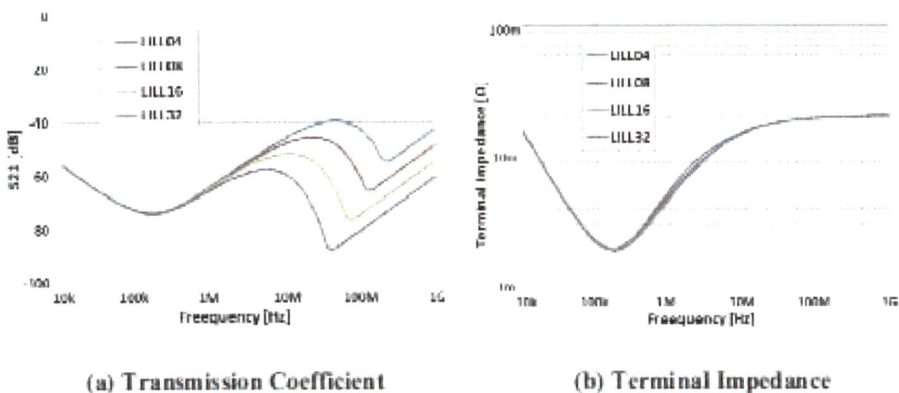

(a) Transmission Coefficient   (b) Terminal Impedance

Figure 8: Characteristics of Design Example of LILL Used on PCB of SMPS.

Fig. **8** shows the calculated transmission coefficient and the terminal impedance when the example of the desirable LILL is used on the PCB. The calculation condition was the following; the size of the power trace to the power receiving terminal was 10mm×200mm, the distance between the power trace and the ground plane was 1.53mm, PCB was made of FR4, and the conventional capacitor of 1mF was connected in parallel, respectively.

Fig. **9** shows the calculated transmission coefficient of an example of MILL for SMPS.

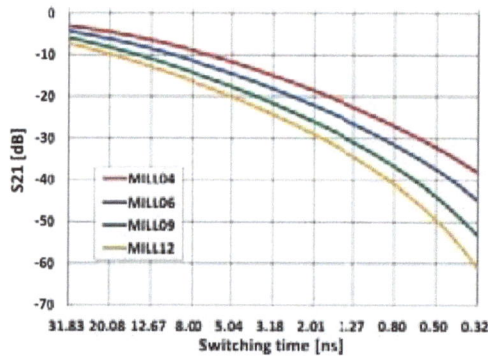

Figure 9: Transmission Coefficient of Design Example of MILL Used on PCB of SMPS

In Fig. **9**, the calculation condition was the following; z was to 12mm from 4mm, C_1 was 1.24μF, R was 0.8, k was 31.7, L was 0.45, w was 2mm, a was 385nm, ρ_C was 300μΩm, each σ of the conductive polymer and the carbon graphite was 12,000S/m and 72,727S/m, μ_r of the alumina was 8.5, b_A for the calculation of the characteristic impedance of the attenuation constant was 0, b_A for the calculation of the characteristic impedance of the reflection coefficient was 0.8μm, b_s was 1μm, b_c was 2.29μm, C_T was $5×10^{-17}F$, the thickness of the etched aluminum film was 115μm, the size of the power trace to the power receiving terminal was 40mm×5mm, the distance between the power trace and the ground plane was 1.53mm, and PCB was made of FR4, respectively. The calculated capacitance of MILL12 of design example is $1.2×10^{-7}F$ and the internal impedance is 1Ω. Therefore, the rise time of the switching voltage is increased to 120ns which is 35times of 3.5ns. The transmission coefficient of MILL12 shown in Fig. **9b** is -23dB or 0.07 at 3.5ns. The calculated result of the maximum wire length will be relaxed to 494 times at between the point D and E. The length of the winding wire of the inductor is not limited when it is used in the magnetic core. The maximum switching frequency is 2.6*MHz*.

Fig. **10** shows the calculated waveform and the shapes on the power line when the length of the power line to LILL from the drain of NMOSFET in Fig. **6** is 500*mm*. The large spike voltage is generated when NMOSFET changes OFF because the large DC current flows on the power line.

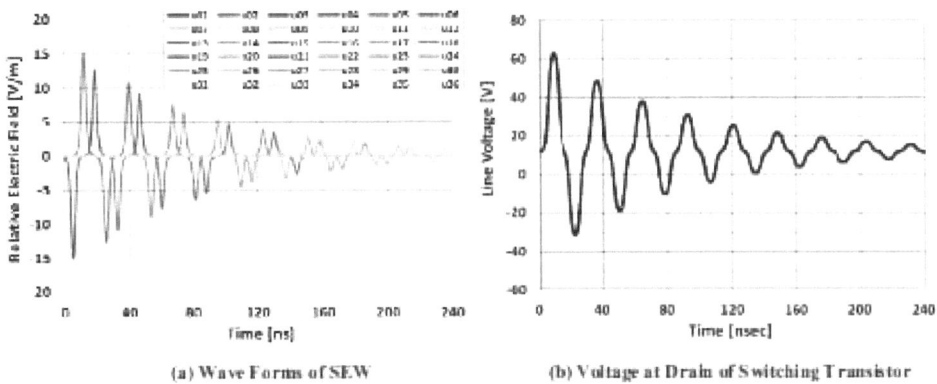

(a) Wave Forms of SEW   (b) Voltage at Drain of Switching Transistor

Figure 10: Calculated Waveform and the Shapes on the Power line Length to LILL is 500mm.

The calculation condition was the following; each width and length of the trace of the power line was 40*mm* and 500*mm*, the thickness of the insulator was 0.108*mm*, and the output impedance of RJK0368DPA was 4.08Ω at MSF of the rise time. In Fig. **10b**, the bounce or the spike having the attenuating resonant frequency and having the large magnitude exists on the power line at long period relatively. The spike voltage of NMOSFET is more than 60*V*. NMOSFET will break down because the absolute maximum voltage of the drain to source of RJK0368DPA is 30*V*.

Fig. **11** shows the calculated waveform and the shapes on the power line forming QSCC. In this example, the line length of the power line to LILL from the drain of NMOSFET is 48*mm* which is the limit length. EMI and the breaking down of NMOSFET will not be caused because the magnitude of the spike and the resonation is small enough.

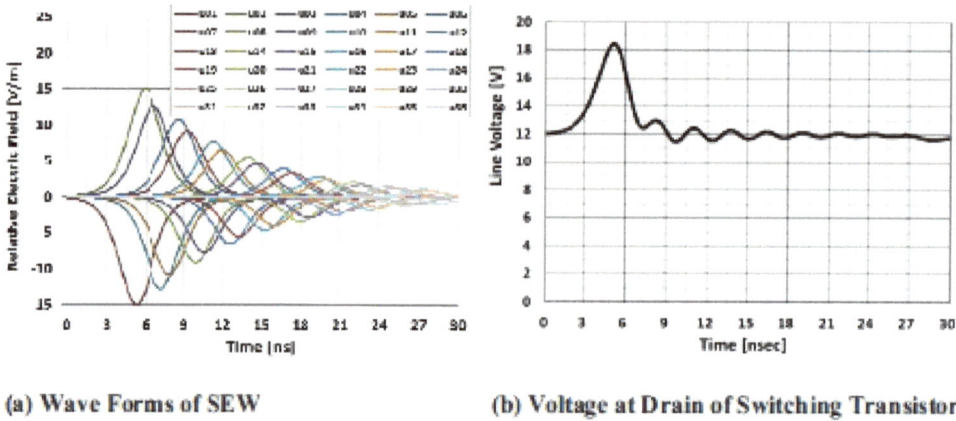

(a) Wave Forms of SEW   (b) Voltage at Drain of Switching Transistor

Figure 11: Calculated Waveform and Shapes when Power Line Length is 48mm.

Fig. 12 shows the calculated waveform and the shapes on the power line in when the length of the power line to LILL from the drain of NMOSFET is 10*mm* which is one fifth approximately of the limit length for forming QSCC. In Fig. 12a and b, only magnitude of SEW is changing and the vibration is not being generated. Therefore, EMI will never be caused.

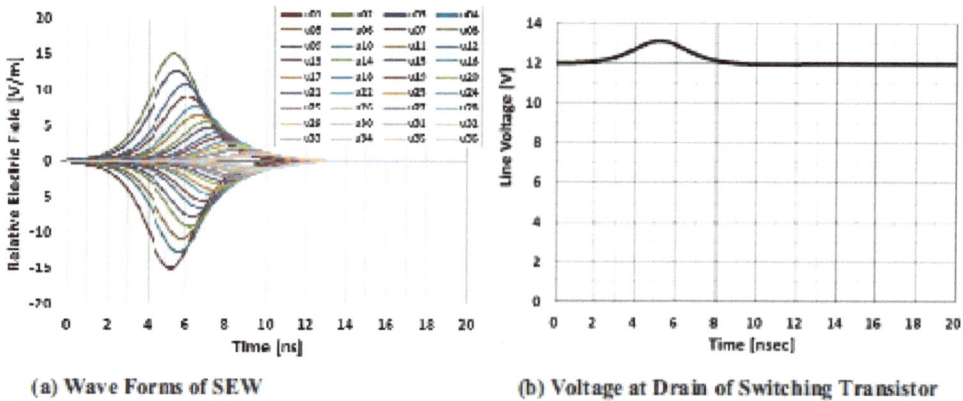

(a) Wave Forms of SEW   (b) Voltage at Drain of Switching Transistor

Figure 12: Calculated Waveform and Shapes when Power Line Length is 10*mm*.

In this way, QSCC of the main switching circuit of the buck converter is formed by using LILL and MILL. When the switching frequency is increased, the length of the chip of LILL and MILL becomes short and the inductor becomes small. The control circuit of the back converter as well should be formed to QSCC in order to form QSCC fully. The control circuit will be reconfigured to QSCC by applying the above mentioned way of the digital circuit.

CONCLUSIONS

When SEMW theory and the technologies of LILL and MILL are applied, SMPS as well as PCB and IC/LSI of the digital circuit can be reconfigured to QSCC. EMI problem does not exist in QSCC. When PCB and IC/LSI of the digital circuit are reconfigured to the complete QSCC, the on-chip clock frequency will be increased to the few hundreds gigahertz and the data transmission will be accelerated greatly till few hundreds bps by using the parallel data transmission instead of the serial data transmission. The improvement of scaling of semiconductor is also necessary for the additional development and the reduction of the power consumption. SPICE can be used to QSCC without problem and the circuit design will become easy because EMI problem does not exist in QSCC. The high-speed serialization/de-serialization (SerDes), the on-chip interconnect formed to the transmission line, and the optical interconnect will become unnecessary when the SMC including the signal line is reconfigured to QSCC. The switching frequency will become over ten megahertz when SMPC is reconfigured to the complete QSCC. However the harmonic waves on the mains cannot be reduced by above mentioned QSCC method because the frequency is too low.

REFERENCES

[1]   H.B. Bakoglu, "Circuits, Interconnections, and Packaging for VLSI", pp. 239-244, Addison-Wesley Pub., 1990.

GLOSSARIES

AC Circuit Theory

It can be say that the history of the electricity started at the discovery of the electrostatic as the lightning by Benjamin Franklin in 1780 after the publication about the magnetostatic by William Gilbert in 1600. The development of the theorems of electromagnetism was started at the publication of the Coulomb's inverse-square law about phenomenon of the electrostatic and the magnetostatic by French physicist Charles-Augustin de Coulomb in 1785. These are the early fruits of the electromagnetism. According to these fruits, the capacitor was invented by Ewald Jürgen Georg von Kleist in 1745, the Ohm's law was presented by Georg Simon Ohm in 1826, and the Kirchhoff's circuit laws which consist of the current law and the voltage law are presented by Gustav Kirchhoff in 1845. The dynamo electric generator for produce the alternating current based on Michael Faraday's principles was constructed by the French instrument maker Hippolyte Pixii in 1832. The alternating current (AC) circuit theory is considered to be established in this age and it has been used for the design and analysis of the AC circuit till now. The direct current (DC) is the unidirectional flow of the electric charge in the conductor, which is called the charge current in as follows. In alternating current (AC), the movement of the electric charge periodically reverses direction. The abbreviations AC and DC are often used to mean simply alternating and direct, as when they modify current or voltage. The usual electric or electronics circuit performs by the DC power supply. The usual AC waveform is generated by the active device in the AC circuit. The passive components such as the resistor, capacitor, inductor, and the transformer are used to reform or transform the AC waveform. The Ohm's law and the Kirchhoff's circuit law for the nodal analysis are used for the design and analysis of the AC circuit. These depend on the lumped element model and the idea that the AC current is the charge current. In the network analysis, 1t is useful to simplify the network by the number of the electric component. In this procedure, the superposition theorem, delta-wye transformation, and the Thévenin's theorem are effective. It is also useful to reconfigure the network to the stereotyped two-terminal network and four-terminal network. In four-terminal network, the admittance parameters are used in the current source AC circuit in usual, the impedance parameters are used in the voltage source AC circuit in usual, the *h* parameters are used for the equivalent circuit of the bipolar transistor in usual, and the *G* parameters are used for the equivalent circuit of the MOEFT or the vacuum tube in usual. The admittance parameters can be got when the source current is given and the circuit voltage was measured, the impedance parameters can be got when the source voltage is given and the circuit voltage was measured, the *h* parameters can be got by varying the condition of four terminals of the bipolar transistor, and the *G* parameters can be got by varying the condition of four terminals of the MOEFT or the vacuum tube.

Fig. **1** shows an example of the equivalent circuit of the NAND gate, which is drown by using the lumped element model.

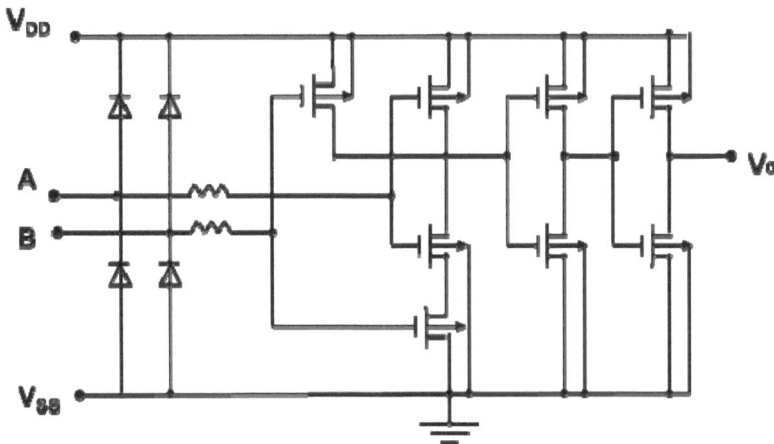

Figure 1: Equivalent Circuit of NAND Gate by using Lumped Element Model.

In Fig. **1**, the size of the devices and the length of the wires are being abbreviated.

From above, the AC circuit theory will be reliable when the electric current consists of only the charge current in the conductor and it cannot handle EMW which has the wave length and travels the outside of the conductors.

Characteristic Equations of Transmission Line

Fig. **2** shows an example of one tiny part of the distributed element model.

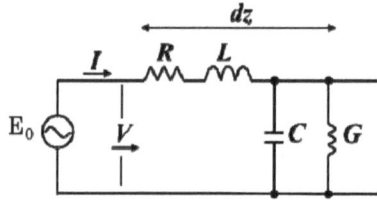

Figure 2: Distributed Element Model.

In Fig. **2**, the actual circuit consists of an infinite series of the distributed element. The distributed element model is applied to the larger circuit than the wavelength of the AC current such as the transmission line. The circuit elements are shown each symbol in accordance with the AC circuit theory. The resistance R means a part of the distributed resistance of the wire. The inductance L means a part of the distributed inductance around the wire. The capacitance C means a part of the distributed capacitance between two wires. The conductance G means a part of the distributed loss of the insulator between two wires. And the electric current is the charge current which flow in the wire.

When the sinusoidal wave travels on the tiny circuit within dz, the magnitude of the voltage (V) and the electric current (I) are:

$$\frac{dV}{dz} = (R + j\omega L)I = ZI \tag{1}$$

$$\frac{dI}{dz} = (G + j\omega C)I = YV \tag{2}$$

From the equation 2:

$$\frac{d^2V}{dz^2} = (R + j\omega L)\frac{dI}{dz} = Z\frac{dI}{dz} \tag{3}$$

By substituting the equation 3 into the equation 2:

$$\frac{d^2V}{dz^2} = ZYV \tag{4}$$

When V is replaced to $e^{\lambda z}$ because the equation 4 is the second degree linear differential equation:

$$\lambda^2 = ZY \tag{5}$$

The equation 5 is called the characteristic equation of the transmission line. From the equation 5:

$$\lambda = \pm\sqrt{ZY} = \pm\sqrt{(R + j\omega L)(G + j\omega C)} = \pm\left(\frac{R\sqrt{C/L}+G\sqrt{L/C}}{2} + j\omega\sqrt{LC}\right) \tag{6}$$

From the equation 6 because the transmission constant is:

$$\gamma = \alpha + j\beta$$

$$\alpha = \frac{R\sqrt{C/L}+G\sqrt{L/C}}{2}, \ \beta = \omega\sqrt{LC} \tag{7}$$

Therefore:

$$V = Ae^{-\gamma z} + Be^{\gamma z} \tag{8}$$

where A and B are the integration constants which are determined by the boundary condition.

As well as the equation 8:

$$I = A\sqrt{\frac{Y}{Z}}e^{-\gamma z} - B\sqrt{\frac{Y}{Z}}e^{\gamma z} \tag{9}$$

From the equation 8 and 9:

$$\frac{V}{I} = \sqrt{\frac{Z}{Y}} \equiv Z_0 \tag{10}$$

where Z_0 is called the characteristic impedance.

The skin effect is defined as the phenomenon of which the charge current approaches to the surface of the conductor at the high-frequency circuit. The skin effect is explained in accordance with the idea of the skin depth defined by the electromagnetism. However the average drift speed of the charge current in the conductor is the order of a millimetre per second and the charge current cannot function in the high-frequency circuit. Therefore, the definition of the skin effect does not correspond to the physics.

From above, the characteristic equations of the transmission line handles the length of the wires, however, it will be reliable when the electric current consists of only the charge current in the conductor and it cannot handle EMW which has the wave length and travels the outside of the conductors.

Charge Current

The charge current is defined to dq/dt, which means the average drift speed of the charge in the conductor. George Gamow first calculated and presented the drift speed of the charge current in the conductor 1947.

The conductor is defined that the condensed state of the positive ion in the gas of the free electron or the charge. The charges flit at the average speed of billions of meters per hour, which corresponds to 1/10 of the light speed approximately. However, the average speed of the charge gas in the free space is zero because the free electron is in the Brownian motion. When the voltage is added to the terminals of the conductive wire, the charge will drift to the positive terminal from the negative terminal. The charge current is the average drift speed of the charge.

The charge current which is the drift value per one second of the summation of the charge in the unit volume of the conductive wire is:

$$I = nevS \tag{11}$$

where n is the summation of the charge per unit volume, e is the charge amount, v is the drift speed of the charge, and S is the square of the cross section of the conductive wire. n is $8.5 \times 10^{28}/m^3$ because n is the value of Avogadro number $(6.02 \times 10^{23}/mol)$ divided by the volume of $1mol$ $(7.1 \times 10^{-6} m^3)$ which is the value of the atomic weight (63.55g) divided by the density $(8.93 \times 10^6 \ g/m^3)$, and e is $-3.2 \times 10^{-19} C$ in the case of the copper wire.

When the charge current of $10A$ flows in the copper wire having the cross section of $1mm^2$, the average speed of the charge or the drift speed of the charge is:

$$v = \frac{I}{neS} = \frac{10}{8.5 \times 10^{28} \times 3.2 \times 10^{-19} \times 10^{-6}} = 3.68 \times 10^{-4} [m/s] \tag{12}$$

From above, it was clarified that the drift speed of the charge current is quite slower than the light speed. Therefore, the charge current can never govern the AC circuit which is the EMW circuit.

Characteristic Impedance of Actual transmission Lines

The characteristic impedance of the parallel plate line is:

$$Z_0 = \frac{h}{w}\sqrt{\frac{\mu_0}{\varepsilon_r \varepsilon_0}} \tag{13}$$

where h is the thickness of the insulator, w is the line width, μ_0 is the permeability in vacuum, ε_0 is the permittivity, and ε_r is the dielectric constant.

The characteristic impedance of the Lecher line is:

$$Z_0 = \sqrt{\frac{\mu_0}{\varepsilon_r \varepsilon_0}}\frac{1}{\pi}\ln\left(\frac{D}{d} + \sqrt{\left(\frac{D}{d}\right)^2 - 1}\right) = \frac{120}{\sqrt{\varepsilon_r}}\cosh^{-1}\left(\frac{D}{d}\right) \tag{14}$$

where d is the diameter of the wirers, D is the distance between wires.

The characteristic impedance of the coaxial line is:

$$Z_0 = \sqrt{\frac{\mu_0}{\varepsilon_r \varepsilon_0}}\frac{1}{2\pi}\ln\left(\frac{D}{d}\right) \tag{15}$$

where d is the diameter of the inner wirer, D is the diameter of the outer conductor.

The equation 15 is effective at the lower frequency than the cut-off frequency which is got by $\pi(d + D)$.

The characteristic impedance of the microstrip line is:

$$Z_0 = \mathrm{k}\cdot\ln\left(1 + \frac{4h}{w_{eff}}\left(\frac{14+\frac{8}{\varepsilon_r}}{11}\frac{4h}{w_{eff}} + \sqrt{\left(\frac{14+\frac{8}{\varepsilon_r}}{11}\frac{4h}{w_{eff}}\right)^2 + \frac{\pi^2}{2}\left(1+\frac{1}{\varepsilon_r}\right)}\right)\right) \tag{16}$$

where w is the width of the strip conductor, t is the thickness of the strip

conductor, $\quad W_{eff} = W + t\dfrac{1+\frac{1}{\varepsilon_r}}{2\pi}\ln\left(\dfrac{4\times2.718}{\sqrt{(t/h)^2+(\pi(w/t+11/10))^{-2}}}\right)$, and k

$= \sqrt{\dfrac{\mu_0}{\varepsilon_0}}\dfrac{1}{2\pi\sqrt{2(1+\varepsilon_r)}}.$

The characteristic impedance of the strip line is:

$$Z_0 = \dfrac{94.15}{\sqrt{\varepsilon_r}\left(\dfrac{W}{h-t}+0.45+1.18\dfrac{t}{h}\right)} \tag{17}$$

where $0.05 \leq t/h \leq 0.5.$

Above mentioned equations for getting the characteristic impedance of the actual transmission lines will be reliable because these are the empirical equations which have been used by many engineers for a long time.

Digital Circuit

Digital circuit is a kind of SMC and it is designed and analysed by the AC circuit theory except the logic design.

Boolean algebra is used for the logical operation basically. The AND gate, the Exclusive OR gate, the NOR gate, the NAND gate, and the amp or the repeater. And the inverters are used to as the primitive circuit in them.

The sequential circuit which consists of the asynchronous sequential circuit and the synchronous sequential circuit and the combinational circuit are used to the digital circuit. The best part of the computer consists of the synchronous sequential circuit. The asynchronous sequential circuit is relatively simple and easy to design but it is susceptible to EMI. The Schmitt trigger is used for improving the immunity to EMI.

One of the most important functions of the digital circuit is the temporary storage of the data. The flip-flop which consists of the set-reset latch, the gated latch, the D flip-flop, the T flip-flop, and the JK flip-flop is used for this object. The gated latch which consists of the gated set-reset latch and the gated D latch is used for the I/O buffer or other data buffering. The data buffering is important for the data processing, the data storage, and the data transmission because the electrical state of all bit which form a data should be stable at the same timing. The difference of the transmission delay is called the skew or the timing skew. The data buffer corrects the timing skew. However, the spending time for the data processing, the data storage, and the data transmission increases when the timing skew is large. The timing skew management is one of the important items for the high-performance digital circuit. However the AC circuit theory and the electromagnetism based on the conventional wave equation are not useful for the skew management.

Dispersion and Group Velocity of the Wave

The carrier wave without the distortion should be modulated when the signal is transmitted by the microwave. The difference of the phase constant (β) or the traveling speed will happen when the microwave is transmitted by the transmission line and the wave guide. As the result, the modulated wave shape which is different from the input wave shape will be observed at the output terminal. In this case, the group of waves is put to traveling situation, and the traveling speed of the wave group is called the group velocity. The group velocity has the frequency characteristics in general. It is said that the transmission line has the dispersibility when the transmitting signal has the broadband spectra because each spectrum travels at the different speed. In this case, the wave will be distorted. It has been believed that the signal wave of digital circuit consists of many harmonics. The signal integrity of the observed wave will be degraded at the output terminal by the dispersion when it is transmitted relatively long distance.

Electric Far Field Strength Radiated by Elementary Dipole

Fig. **3** shows the current distributions on the elementary dipole.

In Fig. **3**, the electric field strength at the far distance radiated from the elementary dipole is:

$$E_\theta = -j \frac{10^{-7} \pi \cdot I_0 \cdot f \cdot L}{r} e^{j(\omega t - \beta r)} \sin \theta \tag{18}$$

where f is the frequency of EMW, I_0 is the antenna current, L is the summation length of the dipole, r is the distance from the doublet to the detecting point, β is $\omega\sqrt{\mu_0\varepsilon_0}$ which is the wave number, ω is $2\pi f$, and θ is the direction of the radiation.

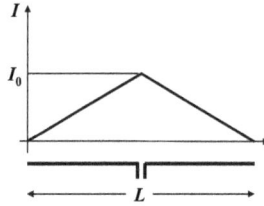

Figure 3: Current Distribution of Elementary Dipole.

In the equation 18 and Fig. **3**, the electric current should not be the charge current explicitly.

Electric Far Field Strength Radiated by Elementary Loop

Fig. **4** shows the magnetic elementary loop. The uniform current flows in the loop conductor.

The electric field strength at the far distance radiated from the elementary loop is:

$$E_\varphi = \frac{10^{-7}4\pi^3\sqrt{\varepsilon_0\mu_0}\cdot I_0\cdot f^2\cdot a^2}{r}e^{j(\omega t-\beta r)}\sin\theta \qquad (19)$$

where a is the radius of the loop, and φ is the direction of the radiation.

In the equation 19, when the uniform current distribution on the circumference is lost, the electric far field strength from the magnetic elementary loop approaches to the characteristics got by the equation 18.

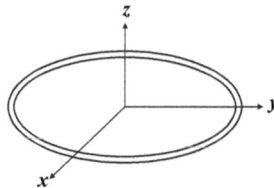

Figure 4: Magnetic Elementary Loop.

In the equation 19 and Fig. **4**, the electric current should not be the charge current explicitly.

EMW Theory [1-3]

The modernized versions of the Maxwell's equations which are formed by four equations were recast by Oliver Heaviside in 1884 from twenty equations which was used in actual for developing the vector wave equations by Maxwell. The Maxwell's equations at the field entirely devoid of matter are:

$$\nabla \times \dot{E} = -\frac{\partial \dot{B}}{\partial t}, \ \nabla \cdot \dot{D} = \rho, \nabla \times \dot{H} = -\frac{\partial \dot{D}}{\partial t} + J, \ \nabla \cdot \dot{B} = 0 \tag{20}$$

where $J = \sigma \dot{E}, \ \dot{D} = \varepsilon_0 \dot{E}, \ \dot{B} = \mu_0 \dot{H}$.

The equation 20 is meaning that the variation of the density of magnetic flux forms the eddy of the electric field having the rotation axis of the vibration direction, the density of electric flux bubbles by the electric charge, the variation of the density of electric flux forms the eddy of the magnetic field having the rotation axis of the vibration direction and the electric current also forms the eddy of the magnetic field, and nothing bubbles the eddy of the magnetic field.

The equation 20 at vacuum entirely devoid of electric charge and current is:

$$\nabla \times \dot{E} = -\mu_0 \frac{\partial \dot{H}}{\partial t}, \ \nabla \cdot \dot{E} = 0, \nabla \times \dot{H} = \varepsilon_0 \frac{\partial \dot{E}}{\partial t}, \ \nabla \cdot \dot{H} = 0 \tag{21}$$

The vector equations arranged the equation 21 is:

$$\nabla^2 \dot{E} - \mu_0 \varepsilon_0 \frac{\partial^2 \dot{E}}{\partial t^2} = 0, \ \nabla^2 \dot{H} - \mu_0 \varepsilon_0 \frac{\partial^2 \dot{H}}{\partial t^2} = 0 \tag{22}$$

The equation 22 was got from above mentioned original twenty equations.

In the equation 22, the propagation speed is got as $1/\sqrt{\mu_0 \varepsilon_0}$ when the equation 22 is compared with the simplest wave equation which was studied by Jean le Rond d'Alembert, *et al.* The value of $1/\sqrt{\mu_0 \varepsilon_0}$ is the light speed. From above, it was clarified that the changing electric field and magnetic form EMW and it propagates at the light speed in vacuum.

When the electric field wave and the magnetic field wave are uniform in the x-y dimension, the arranged equation 22 is:

$$\nabla^2 \dot{E} + \omega^2 \mu_0 \varepsilon_0 \dot{E} = 0, \ \nabla^2 \dot{H} + \omega^2 \mu_0 \varepsilon_0 \dot{H} = 0 \tag{23}$$

A solution of the equation 23 about the vibrating electric field and magnetic field is:

$$E = i\sqrt{2}E_0 \cos\left(\omega\left(t \mp z\sqrt{\mu_0\varepsilon_0}\right) + \theta\right), H = \pm j\sqrt{2}E_0 \sqrt{\frac{\varepsilon_0}{\mu_0}} \cos\left(\omega\left(t \mp z\sqrt{\mu_0\varepsilon_0}\right) + \theta\right) \tag{24}$$

From the equation 24, it can be understood as follows;

a. EMW consists of the orthogonal electric field wave and the magnetic field wave.

b. The magnetic field waves consist of them of both negative and positive magnitude.

c. EMW consist of them of both forward travel and backward travel.

The intrinsic impedance is defined as the electric field divided by the magnetic field. The intrinsic impedance of vacuum is 120π.

In the lossy material, the vector wave equation is:

$$\nabla^2 \dot{E} + \omega^2 \mu\varepsilon\dot{E} - j\omega\mu\sigma\dot{E} = 0, \nabla^2 \dot{H} + \omega^2 \mu\varepsilon\dot{H} - j\omega\mu\sigma\dot{H} = 0 \tag{25}$$

The plane EMW which consists of the electric field and the magnetic field are

$$\frac{\partial^2 \dot{E}_x}{\partial z^2} + (\omega^2 \mu\varepsilon - j\omega\mu\sigma)\dot{E}_x = 0, \frac{\partial^2 \dot{H}_y}{\partial z^2} + (\omega^2 \mu\varepsilon - j\omega\mu\sigma)\dot{H}_y = 0 \tag{26}$$

The reconfiguration of the equation 26 to $\dot{E}_x = Ae^{-\gamma z}$ is:

$$\gamma^2 \dot{A}e^{-\gamma z} + (\omega^2 \mu\varepsilon - j\omega\mu\sigma)\dot{A}e^{-\gamma z} = 0 \tag{27}$$

The recast equation 27 is:

$$\gamma^2 = (\sigma + j\omega\varepsilon)j\omega\mu = \omega^2\mu\varepsilon_0\left(\varepsilon_r - j\frac{\sigma}{\omega\varepsilon_0}\right) \tag{28}$$

In the equation 28, $\varepsilon_r - j\sigma/\omega\varepsilon_0 = e' - je''$ is defined as the complex relative permittivity, and the definition of the loss tangent is:

$$\tan\delta = \frac{\sigma/\omega\epsilon_0}{\varepsilon_r} = \frac{\sigma}{\omega\varepsilon_r\varepsilon_0} \tag{29}$$

The recast equation 28 is:

$$\gamma = \pm\sqrt{(\sigma + j\omega\varepsilon)j\omega\mu} = \pm\omega\sqrt{\mu\varepsilon}\left(1 - j\frac{\sigma}{2\omega\varepsilon} + \frac{1}{8}\left(\frac{\sigma}{\omega\varepsilon}\right)^2 + \cdots\right) \tag{30}$$

From the equation 30, each the real part and the imaginary part is:

$$\alpha \approx \frac{\sigma}{2}\sqrt{\frac{\mu}{\varepsilon}} \tag{31}$$

$$\beta \approx \omega\sqrt{\mu\varepsilon}\left(1 + \frac{1}{8}\left(\frac{\sigma}{\omega\varepsilon}\right)^2\right) \approx \omega\sqrt{\mu\varepsilon} \tag{32}$$

In the equation 30, 31, and 32, γ is called the propagation constant, α is called the attenuation constant, and β is called the phase constant. The unit of the attenuation constant is shown by nep/m which means that EMW attenuates to 0.368 at the traveling of 1m distance.

α of the equation 31 is different from it which is defined in the distributed element model and the telecommunication engineering.

The intrinsic impedance of the lossy material is:

$$Z = \sqrt{\frac{j\omega\mu}{\sigma+j\omega\varepsilon}} \tag{33}$$

A solution about the sinusoidal wave of the equation 24 in the lossy material is:

$$E = i\sqrt{2}E_0\,e^{-\alpha z}\cos(\omega t - \beta z + \theta), H = \pm j\sqrt{2}\frac{E_0}{Z_0}e^{-\alpha z}\cos(\omega t - \beta z + \theta) \tag{34}$$

The propagation constant of the conductor is:

$$\gamma = \alpha + j\beta = \sqrt{(\sigma + j\omega\varepsilon)j\omega\mu} \approx \sqrt{\pi f \mu \sigma} = \sqrt{\frac{\omega\mu\sigma}{2}}(1 + j) \tag{35}$$

Therefore, α and β of the conductor is:

$$\alpha = \beta = \sqrt{\frac{\omega\mu\sigma}{2}} = \sqrt{\pi f \mu \sigma} \tag{36}$$

From the equation 33, the intrinsic impedance of conductor:

$$Z = \sqrt{\frac{j\omega\mu}{\sigma + j\omega\varepsilon}} \approx \sqrt{\frac{j\omega\mu}{\sigma}} = \sqrt{\frac{\omega\mu}{2\sigma}}(1 + j) = \frac{\sqrt{\pi f \mu \sigma}}{\sigma}(1 + j) = \frac{1}{\sigma\delta}(1 + j) \tag{37}$$

In the equation 37, δ is called the skin depth. The skin effect:

The relationship of the electric field and the electric potential or the voltage is:

$$\dot{E} = -\nabla V = -\left(\frac{\partial V}{\partial x}i_x + \frac{\partial V}{\partial y}i_y + \frac{\partial V}{\partial z}i_{zx}\right) \tag{38}$$

From the equation 38, the relationship of the electric field and the voltage between two conductors having the distance r is:

$$\dot{E} = -\frac{\partial V}{\partial r} \quad \text{or} \quad V = -E \cdot r \tag{39}$$

According to the integral form of the original Ampère's circuit law, the relationship of the magnetic field and the electric current about the wire is:

$$I = \oint \dot{H} dl \tag{40}$$

where l is the average length of the magnetic field of the round of the wire.

Four-Terminal Network

Fig. **5** shows circuit of the four-terminal network.

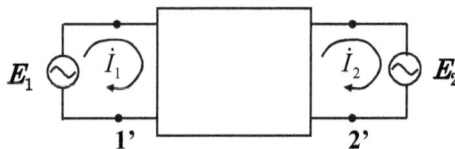

Figure 5: Circuit of Four-Terminal Network.

The circuit equations of the circuit shown in Fig. **5** are:

$$\left.\begin{array}{l}
\dot{Z}_{11}\dot{I}_1 + \dot{Z}_{12}\dot{I}_2 + \dot{Z}_{13}\dot{I}_3 + \cdots + \dot{Z}_{1n}\dot{I}_n = \dot{E}_1 \\[6pt]
\dot{Z}_{21}\dot{I}_1 + \dot{Z}_{22}\dot{I}_2 + \dot{Z}_{23}\dot{I}_3 + \cdots + \dot{Z}_{21}\dot{I}_n = -\dot{E}_1 \\[6pt]
\dot{Z}_{31}\dot{I}_1 + \dot{Z}_{32}\dot{I}_2 + \dot{Z}_{33}\dot{I}_3 + \cdots + \dot{Z}_{31}\dot{I}_n = 0 \\[6pt]
\vdots\ \vdots\ \vdots\ \vdots\ \vdots\ \vdots \\[6pt]
\dot{Z}_{n1}\dot{I}_1 + \dot{Z}_{n2}\dot{I}_2 + \dot{Z}_{n3}\dot{I}_3 + \cdots + \dot{Z}_{nn}\dot{I}_n = 0
\end{array}\right\} \tag{41}$$

The solutions of the equation 6 about the current are:

$$\dot{I}_1 = \frac{\dot{\Delta}_{11}\dot{E}_1 - \dot{\Delta}_{21}\dot{E}_2}{\dot{\Delta}}, \quad \dot{I}_2 = \frac{\dot{\Delta}_{12}\dot{E}_1 - \dot{\Delta}_{22}\dot{E}_2}{\dot{\Delta}} \tag{42}$$

Where,

$$\dot{\Delta} = \begin{vmatrix} \dot{Z}_{11} & \dot{Z}_{12} & \cdots & \dot{Z}_{1n} \\ \dot{Z}_{21} & \dot{Z}_{22} & \cdots & \dot{Z}_{2n} \\ \vdots & \vdots & \vdots & \\ \dot{Z}_{n1} & \dot{Z}_{n2} & \cdots & \dot{Z}_{nn} \end{vmatrix}, \quad
\dot{\Delta}_{11} = \begin{vmatrix} \dot{Z}_{22} & \dot{Z}_{23} & \cdots & \dot{Z}_{2n} \\ \dot{Z}_{32} & \dot{Z}_{33} & \cdots & \dot{Z}_{3n} \\ \vdots & \vdots & \vdots & \\ \dot{Z}_{n2} & \dot{Z}_{n3} & \cdots & \dot{Z}_{nn} \end{vmatrix}, \quad
\dot{\Delta}_{12} = \begin{vmatrix} \dot{Z}_{21} & \dot{Z}_{23} & \cdots & \dot{Z}_{2n} \\ \dot{Z}_{31} & \dot{Z}_{33} & \cdots & \dot{Z}_{3n} \\ \vdots & \vdots & \vdots & \\ \dot{Z}_{n1} & \dot{Z}_{n3} & \cdots & \dot{Z}_{nn} \end{vmatrix}, \text{ and}$$

$$\dot{\Delta}_{22} = \begin{vmatrix} \dot{Z}_{11} & \dot{Z}_{13} & \cdots & \dot{Z}_{1n} \\ \dot{Z}_{31} & \dot{Z}_{33} & \cdots & \dot{Z}_{3n} \\ \vdots & \vdots & \vdots & \\ \dot{Z}_{n1} & \dot{Z}_{n3} & \cdots & \dot{Z}_{nn} \end{vmatrix}.$$

The equation 6 becomes:

$$\dot{I}_1 = \dot{Y}_{11}\dot{E}_1 - \dot{Y}_{12}\dot{E}_2, \quad \dot{I}_2 = \dot{Y}_{21}\dot{E}_1 - \dot{Y}_{22}\dot{E}_2 \tag{43}$$

where $\dot{Y}_{11} = \frac{\dot{\Delta}_{11}}{\dot{\Delta}}, \dot{Y}_{12} = \frac{\dot{\Delta}_{21}}{\dot{\Delta}}, \dot{Y}_{21} = \frac{\dot{\Delta}_{12}}{\dot{\Delta}}, \dot{Y}_{22} = \frac{\dot{\Delta}_{22}}{\dot{\Delta}}$

The matrix equation of the equation 42 is:

$$\begin{vmatrix} \dot{I}_1 \\ \dot{I}_2 \end{vmatrix} = \begin{vmatrix} \dot{Y}_{11} & \dot{Y}_{12} \\ \dot{Y}_{21} & \dot{Y}_{22} \end{vmatrix} \cdot \begin{vmatrix} \dot{E}_1 \\ -\dot{E}_2 \end{vmatrix} \tag{44}$$

where each $\dot{Y}_{11}$, $\dot{Y}_{12}$, $\dot{Y}_{21}$, and $\dot{Y}_{22}$ is called the admittance parameter.

From the equation 44, the matrix equation about the impedance parameter is:

$$\begin{vmatrix} \dot{E}_1 \\ -\dot{E}_2 \end{vmatrix} = \begin{vmatrix} \dot{Z}_{11} & \dot{Z}_{12} \\ \dot{Z}_{21} & \dot{Z}_{22} \end{vmatrix} \cdot \begin{vmatrix} \dot{I}_1 \\ \dot{I}_2 \end{vmatrix} \tag{45}$$

where $\dot{Z}_{11} = \frac{\dot{Y}_{22}}{\dot{\Delta}_y}$, $\dot{Z}_{12} = \frac{\dot{Y}_{12}}{\dot{\Delta}_y}$, $\dot{Z}_{21} = \frac{\dot{Y}_{21}}{\dot{\Delta}_y}$, $\dot{Z}_{22} = \frac{\dot{Y}_{11}}{\dot{\Delta}_y}$ and $\dot{\Delta}_y = \dot{Y}_{12}\dot{Y}_{22} - \dot{Y}_{12}\dot{Y}_{21}$.

From the equation 45, the matrix equation about the *h* parameter is:

$$\begin{vmatrix} \dot{E}_1 \\ \dot{I}_2 \end{vmatrix} = \begin{vmatrix} \dot{H}_{11} & \dot{H}_{12} \\ \dot{H}_{21} & \dot{H}_{22} \end{vmatrix} \cdot \begin{vmatrix} \dot{I}_1 \\ -\dot{E}_2 \end{vmatrix} \tag{46}$$

From the equation 45, the matrix equation about the *G* parameter is:

$$\begin{vmatrix} \dot{I}_1 \\ -\dot{E}_2 \end{vmatrix} = \begin{vmatrix} \dot{G}_{11} & \dot{G}_{12} \\ \dot{G}_{21} & \dot{G}_{22} \end{vmatrix} \cdot \begin{vmatrix} \dot{E}_1 \\ \dot{I}_2 \end{vmatrix} \tag{47}$$

From above, it can say that above equations are true when the electric current consists of only the charge current.

Fourier Transform

The usual waveform of an AC circuit is a sine wave. In certain applications, different waveforms are used, such as triangular or square waves. They are called the distorted waveform. The distorted waveform can be transformed by the Fourier Transform to many harmonic waves of which shape is the sinusoidal. The Fourier Transform was presented in first as the Fourier series in mathematics by Joseph Fourier in 1807. It has been considered that SMC can be designed and analyzed by the AC circuit theory inclosing the Fourier Transform. The Fourier transform is often used to the EMI analysis because the time domain can be transformed to the frequency domain, and *vice versa*.

When the function y(x) of the variable x is vibrated with having the period of 2π, which is shown by $y(x) = y(x + 2\pi)$:

$$y(x) = a_1 \sin x + a_2 \sin 2x + \cdots\cdots + b_0 + b_1 \cos x + b_2 \cos 2x + \cdots\cdots$$

$$= \sum_{n=1}^{\infty} a_n \sin nx + b_0 + \sum_{n=1}^{\infty} b_n \cos nx \qquad (48)$$

where n is integer number, $a_n = \frac{1}{\pi}\int_0^{2\pi} y(x) \sin nx \, dx$, $b_0 = \frac{1}{2\pi}\int_0^{2\pi} y(x) \, dx$, $b_n = \frac{1}{\pi}\int_0^{2\pi} y(x) \cos nx \, dx$.

When the function has the period T, $y(t)$ can be expressed in $y(t+T)$.

When the function $y(t)$ is distorted wave,

$$y(t) = \sum_{n=1}^{\infty} a_n \sin n\omega t + b_0 + \sum_{n=1}^{\infty} c_n \cos n\omega t \qquad (49)$$

In the equation 48, when $c_n = \sqrt{a_n{}^2 + b_n{}^2}$ and $tan^{-1}\frac{a_n}{b_n} = \theta_n$

$$y(t) = b_0 + c_1 \sin(\omega t + \theta_1) + c_2 \sin(2\omega t + \theta_2) + \cdots\cdots \qquad (50)$$

In the equation 49, $c_1 \sin(\omega t + \theta_1)$ is called the function wave and $c_2 \sin(2\omega t + \theta_2)$ is called the second harmonic. They are called the higher harmonics or harmonics.

Fig. **6** shows the rectangular voltage shape.

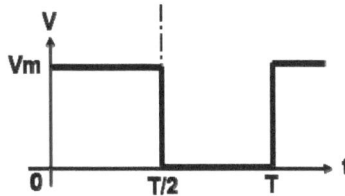

Figure 6: Rectangular Voltage Shape.

The Fourier expansion of the rectangular voltage shape shown in Fig. **6** is,

$$y(t) = \frac{4V_m}{t}\left(\sin \omega t + \frac{1}{3}\sin 3\omega t + \frac{1}{5}\sin 5\omega t + \cdots\cdots\right) \qquad (51)$$

The triangular shape, trapezoidal shape, and the one-shot pulse shape can be transformed to the gathering of the sinusoidal waves by the Fourier expansion.

Fig. **7** shows the example of the trapezoidal voltage shape.

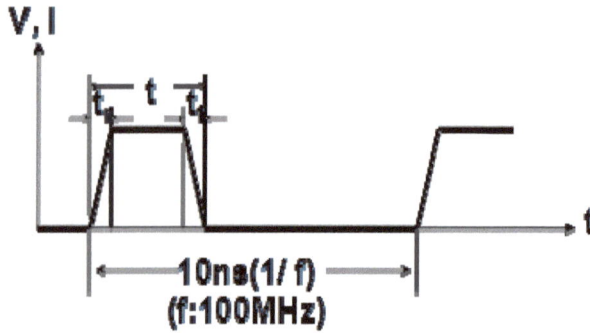

Figure 7: Example of Trapezoidal Voltage Shape.

Fig. **8** shows the calculated spectra of the trapezoidal voltage shape by the Fourier expansion. The calculation condition of the spectra shown in Fig. **8a** was the following; the magnitude is 1, the rise time t_r and the fall time t_f are 100ps, and the pulse width t is 0.5ns, 1ns, and 4.9ns. The calculation condition of the spectra shown in Fig. **8b** is the following; the magnitude is 1, the rise time t_r and the fall time t_f are 10ps, and the pulse width t is 1fs, 0.1ns, 0.5ns, 1ns, and 4.99ns.

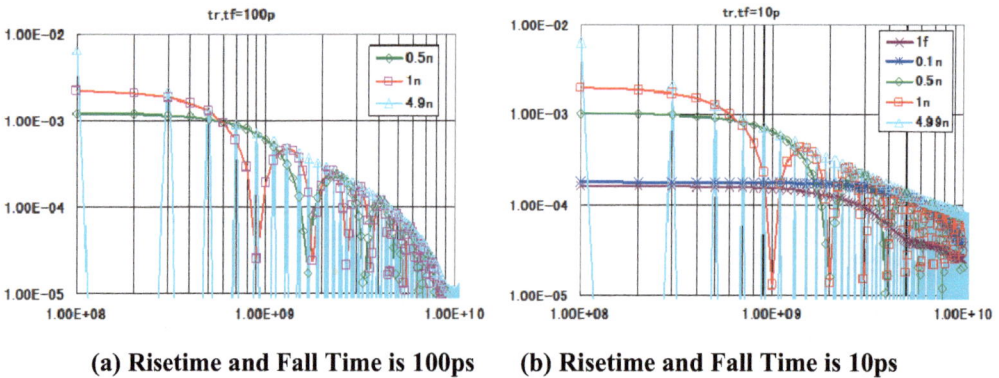

(a) Risetime and Fall Time is 100ps   **(b) Risetime and Fall Time is 10ps**

Figure 8: Calculated Spectra of Trapezoidal Voltage Shape.

In Fig. **8**, b_0 in the equation 18 is not shown. According to the equation 18, b_0 depends on the duty cycle. Therefore, b_0 is DC which means the stationary state. However, the stationary state of the trapezoidal voltage shape shown in Fig. **7** exists at the high level and the low level explicitly.

From above, the Fourier Transform is not reliable when it is applied to the SMC, because the exiting states and the stationary state are not considered and it is not in accordance with physics.

Kirchhoff's Circuit Laws

Fig. **9** shows the description schematics of the Kirchhoff's circuit laws.

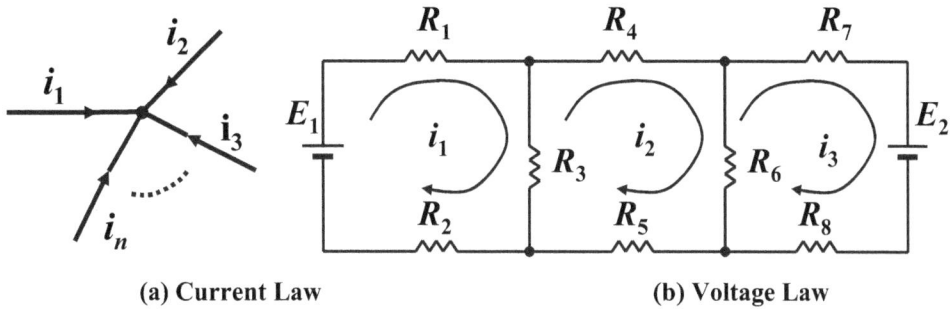

(a) Current Law　　　　　　　　　(b) Voltage Law

Figure 9: Description Schematics of Kirchhoff's Circuit Laws.

In Fig. **9a**, algebraic sum total of the influent current of each point is zero. Therefore,

$$\sum_{j=1}^{n} i_j = 0 \tag{52}$$

In Fig. **9b**, the sum total of the voltage drops is equal to the sum total of the electromotive forces. The circuit equations of Fig. **1b** are:

$$(R_1 + R_2)i_1 + R_3(i_1 - i_2) = E_1 \tag{53}$$

$$R_3(i_2 - i_1) + (R_4 + R_5)i_2 + R_6(i_2 - i_3) = 0 \tag{54}$$

$$(R_7 + R_8)i_3 + R_6(i_3 - i_2) = E_2 \tag{55}$$

The electric current i_1, i_2, and i_3 can be got by solving the equations of 53, 54, and 55.

From above, it can say that the Kirchhoff's circuit laws are true when the electric current consists of only the charge current.

Laplace Transform

The Laplace transform is used for the transient analysis of SMC as well as the circuit of the switching mode power supply. The Laplace transform is related to the Fourier Transform, but whereas the Fourier transform expresses the continuous phenomenon, the Laplace transform resolves the transient phenomenon of the wave. Like the Fourier transform, the Laplace transform is used for solving differential and integral equations. In analysis of the transient phenomenon of SMC, the circuit equation is formed in accordance with the AC circuit theory at first. Next the functions of the time domain in the circuit equation are transformed to the Laplace s-domain. Next the current and the voltage are solved. And the functions of the time domain is got by the inverse Laplace transform finally. The tables of the selected Laplace transform and inverse Laplace transform of the circuit networks are used at the actual circuit design [4]. The Laplace transform is used for the transient analysis of the SMPS circuit usually because many types of the circuit component and device such as the inductor, capacitor, resistor, transformer, diode, and MOSFET are used.

From above, it can say that the Laplace transform is true when the electric current consists of only the charge current.

Lumped Element Model

The placement of the circuit element of the AC circuit can be confirmed usually. Such circuit consists of the AC circuit theory and it is called the lumped element model. In the lumped element model makes simplifying assumption that the characteristics of resistance, capacitance, and reactance are concentrated into the idealized electrical component such as resistors, capacitors, inductors, and the others joined by the network of the conductive wires. This technique can be applied to the relatively smaller circuit than the wavelength of the AC current.

Non Linear Undulation Theory [5, 6]

The solitary wave was found out as the great wave by John Scott Russell who was making an experiment for smoothing the run of the boat in a canal in 1834. However this discovery was not recognized by the scientist at this time.

D.J. Korteweg and G. deVries presented the equation about the wave traveling one direction on the shallow water in 1895. Developed nonlinear evolution equation which governs the surface gravity waves propagating in the shallow channel water having the long one-dimensional and small amplitude is:

$$\frac{\partial \eta}{\partial \tau} = \frac{3}{2}\sqrt{\frac{g}{h}} \cdot \frac{\partial}{\partial \xi}\left(\frac{1}{2}\eta^2 + \frac{2}{3}\alpha\eta + \frac{1}{3}\sigma\frac{\partial^2\eta}{\partial \xi^2}\right) \tag{56}$$

where $\frac{1}{3}h^3 - Th/(\rho g)$, η is the surface evaluation above the equilibrium level h, α an small arbitrary constant related to the uniform motion of the liquid, g the gravitation constant, T the surface tension and ρ the density (the terms "long" and "small" are meant in comparison to the depth of the channel). The controversy was now resolved since equation 56, now known as the Korteweg-deVries (KdV) equation, has permanent wave solutions, including the solitary wave solutions. When

$$t = \frac{1}{2}\sqrt{g/(h\sigma)}\tau, x = -\sigma^{-\frac{1}{2}}\xi, u = \frac{1}{2}\eta + \frac{1}{3}\alpha \tag{57}$$

The equation 56 becomes:

$$u_t + 6uu_x + u_{xxx} = 0 \text{ or } du/dt = u_t + uu_x \tag{58}$$

where subscripts denote partial differentiations.

A solution of the KdV equation 56 about the wave which has the quantity is:

$$u(x.t) = 2\kappa^2 sech^2\{\kappa(x - 4\kappa^2 t - x_0)\} \tag{59}$$

where κ and x_0 are constants, $4\kappa^2$ is the velocity of this wave and it is proportional to the amplitude of $2\kappa^2$.

Therefore, taller wave travel faster than shorter ones.

In 1965, N.J. Zabusky and M.D. Kruskal discovered that these solitary wave solutions have the remarkable property that the interaction of two solitary wave solutions is elastic, and are called soliton.

Fig. **10** shows an example of the traveling wave form of the soliton.

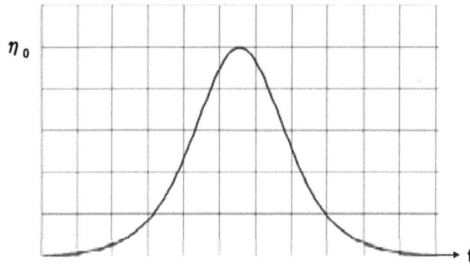

Figure 10: Setup for Measuring S Parameters of DUT.

In Fig. **10**, the waveform of the soliton is similar to the Gaussian wave. The soliton is the single wave but the Gaussian wave consists of many harmonic waves. Therefore the Gaussian wave does not maintain the waveform at the propagation in the lossy material.

The confirmed characteristics of the soliton are the following;

a. The soliton travels with keeping its speed and waveform.

b. The magnitude attenuates at the propagation in the lossy material.

c. The magnitude and the speed of the soliton are not influenced by crossing.

d. When the magnitude becomes larger, the wave length becomes smaller and the propagation speed becomes faster. However the propagation speed of the massless wave is constant.

Parity

The parity check circuit is used for detecting the error caused by the electromagnetic disturbance. Therefore, the parity check circuit is not effective to the continuous disturbance such as the crosstalk, power noise, and the substrate noise, but effective the intermittent disturbance.

A parity bit is a bit that is added to ensure that the number of bits with the value one in a set of bits is even or odd. Parity bits are used as the simplest form of error detecting code. When the error is detected by the parity check circuit, the data processing is retried. The parity is used to SCSI bus and OCI bus for example in the personal computer and used to many microprocessors. In the serial data transmission which avoids the serious error from the crosstalk, the common format is seven data-bit, one even parity bit, and one or two stop bit. Recovery from the error is usually done by re-transmitting the data, the details of which are usually handled by software.

Reflection Coefficient

When the transmission line of which characteristic impedance is Z_0 is connected to the load (Z_L) and the step voltage source, the voltage and the current on the transmission line are:

$$V_i = Z_0 I_i, \ V_t = Z_L I_t \tag{60}$$

where each V_i and I_i is the voltage and the current on the transmission line, each and V_t and I_t is the voltage and the current of the load.

In the equation 60, when V_i and I_i are equal to V_t and I_t, the signal power is absorbed in the load perfectly. This state is called the state of the impedance matching.

When the load impedance is mismatching to the characteristic of the transmission line, the reflection voltage is:

$$V_r = \frac{Z_L - Z_0}{Z_L + Z_0} V_1 = \Gamma_L \tag{61}$$

where V_r is the reflection voltage and V_1 is the supply voltage and a_L is called the reflection coefficient.

The reflection generates the bounce or the ringing on the transmission line. The lattice diagram, the bounce diagram, the echo diagram, or the flection diagram has been used for analyzing the reflection phenomenon on the transmission line [7, 8].

Scattering Matrix

EMW which has the transverse electromagnetic (TEM) mode is effective on the actual transmission line which consists of two isolated conductors. Although, the wave guide which consist of one conductor tube is also used as the transmission line. EMW consists of the TEM mode and other mode. When the signal is not the TEM mode EMW or non-TEM mode EMW is mixed, the telegrapher's equations and the characteristic impedance become void. In such case, the magnitude of the traveling EMW should be handled by the power.

The scattering matrix about a_n and b_n is:

$$[b] = [S][a] \quad \text{or} \quad \begin{bmatrix} b_1 \\ \vdots \\ b_n \end{bmatrix} = \begin{bmatrix} S_{11} & \cdots & S_{1n} \\ \vdots & \cdots & \vdots \\ S_{n1} & \cdots & S_{nn} \end{bmatrix} \cdot \begin{bmatrix} a_1 \\ \vdots \\ a_n \end{bmatrix} \tag{62}$$

In the equation 61, each S_{nn} is called the S parameter, each a_n and b_n is called the wave amplitude which is defined as the square of the power, a_n is the wave amplitude of the injection wave, b_n is the wave amplitude of the reflection wave, and the phase is the complex number which is the phase of the transverse component of the electric field,

Fig. **11** shows the setup for measuring the S parameter of DUT.

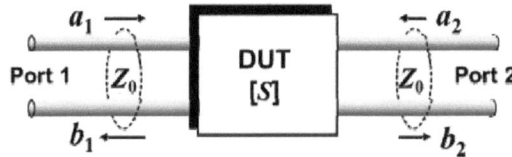

Figure 11: Setup for Measuring S Parameters of DUT.

In Fig. **11**, the port 1 and the port 2 is connected to the network analyzer by the transmission lines which have the characteristic impedance Z_0, The value of Z_0 and is 50Ω usually. The measured S_{11}, S_{22} correspond to the reflection coefficient of DUT, and the measured S_{21}, S_{12} correspond to the transmission coefficient of DUT.

When the radiation loss from the transmission line is negligible, the relationship between the reflection coefficient and the transmission coefficient at the connection boundary of two transmission lines is:

$$S_{11}^{2} + S_{21}^{2} = 1 \tag{63}$$

When the radiation loss and the absorption loss exist, the equation 63 becomes:

$$|S_{11}|^{2} + |S_{21}|^{2} + \frac{P_{rad}+P_{abs}}{P_{inc}} = 1 \tag{64}$$

where P_{rad} is the radiation loss, P_{abs} is the absorption loss, and P_{inc} is the incoming power.

The scattering matrix is reliable because it handles the power of EMW.

Significant Frequency

The significant frequency (SF) of the trapezoidal pulse is defined as $0.34/t_r$, where t_r is the ramp time of the pulse [9].

Properties are the following;

    a.   For each trapezoidal pulse, about 15% of its frequency component or harmonic waves are at frequencies higher than the SF.

    b.   The magnitude of the pulse's spectrum at a frequency higher than the SF is less than 10% of its maximum value. Beyond the SF, the spectral amplitude rolls off faster than the 20 dB/decade.

These two properties indicate that although 15% of the frequency components of the pulse are at a higher frequency than SF, the overall magnitude of them is small. Therefore, SF is a very good representative of a trapezoidal pulse for approximating its high-frequency electrical behavior, especially when deciding whether inductance needs to be considered.

The idea of SF cannot be applied to the electromagnetic analysis of SMC directly because it depends on the Fourier Transform.

SPICE

SPICE is the most famous circuit simulator based on the AC circuit theory. SPICE which was presented in 1971 has been used for the actual design and analyses of almost all AC circuit. IC and LSI consist of the microscopic circuit and the measurement of the circuit voltage and current is quite difficult. Therefore, the spice has been used happen to the design and analysis and it has been improved through this application. In recent, SPICE became used to the analysis of EMI and the signal integrity of the SMC by using the sub-circuit effectively.

Telegrapher's Equations

The principle theorem of the telecommunication engineering is the telegrapher's equations. The telegrapher's equations consist of a pair of linear differential equations which are described by the voltage and current on the transmission line with distance and time. The equations were developed by Oliver Heaviside in 1880s for the telecommunication engineering. The theory has been applied to the transmission lines of all frequencies including high-frequency transmission lines, low frequency such as power lines.

Fig. **12** shows the distributed element model of the transmission line.

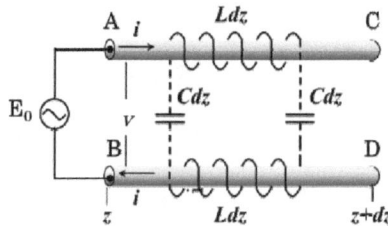

Figure 12: Distributed Element Model of Transmission Line.

In Fig. **12**, i is the charge current in the conductor, dz is the differential length, and the elements R and G are neglected because they of the actual transmission line are negligibly small.

The circuit equations of Fig. **4** are:

$$Ldz\frac{\partial i}{\partial t} = dV, Cdz\frac{\partial V}{\partial t} = \frac{\partial V}{\partial t} = di \tag{65}$$

The transformed equations which are called the telegrapher's equations are:

$$L\frac{\partial i}{\partial t} = \frac{\partial V}{\partial z}, \quad C\frac{\partial V}{\partial t} = \frac{\partial i}{\partial z} \tag{66}$$

The telegrapher's equations become:

$$L\frac{\partial^2 i}{\partial t^2} = \frac{\partial^2 V}{\partial t \partial z}, C\frac{\partial^2 V}{\partial t \partial z} = \frac{\partial^2 i}{\partial z^2} \tag{67}$$

From the equation 67, the equation about the charge current is:

$$LC\frac{\partial^2 i}{\partial t^2} = \frac{\partial^2 i}{\partial z^2} \tag{68}$$

The propagation speed c is got as $c = 1/\sqrt{LC}$ when the equation 68 is compared with the simplest wave equation which was studied by Jean le Rond d'Alembert, *et al.* In Fig. **4**, when the insulator consists of vacuum, c becomes the light speed.

From above, it can say that the telegrapher's equations are true when the electric current consists of only the charge current.

Vector Wave Equations

The vector wave equations are an important second-order linear partial differential equation for the description of waves such as sound waves, light waves and water waves. It arises in fields like acoustics, electromagnetics, and fluid dynamics. Historically, the problem of a vibrating string such as that of a musical instrument was studied by Jean le Rond d'Alembert, Leonhard Euler, Daniel Bernoulli, and Joseph-Louis Lagrange.

The simplest linear wave equation which is shown by the hyperbolic partial differential equations is:

$$\frac{1}{c^2}\frac{\partial^2 u}{\partial t^2} = \nabla^2 u \tag{69}$$

where c is the propagation speed of the wave and $\nabla^2 = \frac{\partial^2}{\partial z_1^2} + \frac{\partial^2}{\partial z_2^2} + \cdots \frac{\partial^2}{\partial z_n^2}$ which is Laplacian.

The result is the d'Alembert's formula as the solution of the equation 69 is:

$$u(x,t) = \frac{f(x-t)+f(x+t)}{2} + \frac{1}{2c}\int_{x-ct}^{x+ct} g(s)ds \tag{70}$$

In equation 70, $f(x\text{-}t)$ means a linear wave which travels to x direction, and $f(x\text{+}t)$ means another linear wave which travels to $-x$ direction.

REFERENCES

[1]   W.K.H. Panofsky, "Classical Electricity and Magnetism: Second Edition", Dover Publications, Inc., 2005.

[2]   J.A. Atratton, "Electromagnetic Theory", IEEE press, Wiley-Interscience, AJohn & Sons Inc., 2010.

[3]   L Solymar, "Lectures on Electromagnetic Theory, A short Cource for Engineers", Oxford Univercity press, 1976.

[4]   J.G. Holbrook, "laplace transforms for electonic enguneers", pergamon press limited, New York, 1959.

[5]   G. L. Lamb, Jr, "Element of Soliton Theory", Jhon Wiley, New York, 1980.

[6]   M. J. Ablowitz and P. A. Clarkson, "Soliton, Nonlinear Evolution Equations and Inverse Scattering", Cambridge University Press, London Mathematical Society Lecture Note Series 149, Cambridge 1992.

[7]   Trueman, C.W., "Animating transmission-line transients with bounce", IEEE, Transactions on Education, pp. 115 – 123, Vol. 46, Issue 1, 2003.

[8]   C.-T. Tsai, "Signal integrity analysis of high-speed, high-pin-count digital packages", IEEE, Electronic Components and Technology Conference, vol.2, pp. 1098 – 1107, 1990.

[9]   C.-K. Cheng, *et al.*, "Interconnect Analysis and Synthesis, John Wiley & Sons, INC., pp. 107- 108, 2000.

Index

www.ingramcontent.com/pod-product-compliance
Lightning Source LLC
Chambersburg PA
CBHW050825220326
41598CB00006B/311